Fundamentals of Steelmaking Metallurgy

Fundamentals of Steelmaking Metallurgy

Brahma Deo

Professor, Department of Metallurgical Engineering,
Indian Institute of Technology, Kanpur, India

Rob Boom

Manager Research Steel Division
and
Deputy Head, Corporate Research Laboratorium,
Hoogovens Groep BV, IJmuiden, The Netherlands

Prentice Hall International

New York   London   Toronto   Sydney   Tokyo   Singapore

First published 1993 by
Prentice Hall International (UK) Limited
Campus 400, Maylands Avenue
Hemel Hempstead
Hertfordshire, HP2 7EZ
A division of
Simon & Schuster International Group

Typeset in 10/12 pt Times
by P&R Typesetters Ltd, Salisbury, UK

Printed and bound in Great Britain
by Bookcraft, Midsomer Norton

Library of Congress Cataloging-in-Publication Data

Deo, Brahma
    Fundamentals of steelmaking metallurgy/Brahma Deo & Rob Boom.
      p.   cm.
    Includes bibliographical references and index.
    ISBN 0–13–345380–4
    1. Steel—metallurgy.   I. Boom, Rob.   II. Title.
TN730.D46 1993
669'.142—dc20   92–38513
                                                    CIP

British Library Cataloguing in Publication Data

A catalogue record for this book is available from
the British Library

ISBN 0-13-345380-4 (hbk)

1 2 3 4 5 97 96 95 94 93

Contents

v

Preface

The literature on fundamentals of steelmaking metallurgy is scattered in review articles, research papers and books on process metallurgy. We felt that a book focused on the fundamentals of contemporary steelmaking processes would be of great help to undergraduate students of metallurgical engineering and also to graduates and research engineers working in steel industries all over the world.

The first chapter, on physical fundamentals, deals with the physical interaction between gas bubbles and liquid metal. The behaviour of submerged and impinging gas jets, which form an important part of modern steelmaking methods, are described.

Chapter 2, on thermodynamic fundamentals, covers all necessary aspects of the thermodynamics of dilute liquid iron alloys and also the physical chemistry and thermodynamics of steelmaking slags. Equilibrium calculations in multicomponent multiphase systems, typical of steelmaking, are explained and then applied to the deoxidation of steel.

Chapter 3, on kinetic fundamentals, makes use of the material discussed in the previous two chapters. To start with, gas–slag, gas–metal and slag–metal reactions are discussed separately. Finally, the complex slag–metal–gas reactions are dealt with as combinations of two-phase reactions.

A very brief description of various processes and process routes in primary and secondary steelmaking is given in Chapter 4. The metallurgical aspects of slag formation in primary steelmaking, including metal droplet generation due to jet impingement and the phenomenon of slag foaming, are discussed in Chapter 5. This is followed, in Chapter 6, by a discussion of the kinetics of chemical reactions in primary steelmaking; a dynamic model of the oxygen steelmaking process is built up from first principles.

The metallurgical fundamentals of various treatments in secondary steelmaking, including the effect of stirring and vacuum on slag–metal–gas reactions, are discussed in Chapter 7. The powdered injection of alloys and refining fluxes into liquid steel, although a part of the secondary steelmaking operation, is discussed separately in Chapter 8 as a special topic, namely 'injection metallurgy'. This is because injection metallurgy also incorporates the pretreatment of hot metal produced from blast furnaces.

Several numerical examples are worked out in each chapter to assist in a clearer understanding of the fundamental concepts involved.

The material presented in this book should form a fundamental base for understanding all the existing processes in primary and secondary steelmaking.

Although, fundamentally, the reacting slag, metal and gas phases have remained the same since the Iron Age, new developments have stemmed essentially from a clear understanding and clever application of the fundamental concepts. We hope that this book on fundamentals will help to provide the much needed crucial insight for future developments and innovations in the steel industry.

Brahma Deo
Rob Boom

Acknowledgements

We are grateful to Ir W. Hoenselaar, Director of Technology, Hoogovens IJmuiden, for his kind permission to publish this book; without encouragement and support from Hoogovens IJmuiden, this book could not have been written in its present form.

We are also grateful to The Director, Indian Institute of Technology (IIT), Kanpur, India, for providing all the necessary support and facilities at Kanpur and for granting long leave to one of us (BD) during May–December 1988, May–July 1989 and December 1989–July 1991 to work at Hoogovens IJmuiden as a Consultant and Visiting Professor.

Most of the material in this book is based on the lecture notes prepared for the standard courses on thermodynamics, kinetics and steelmaking metallurgy taught at the IIT, Kanpur, during 1983–92. A part of these lecture notes was updated and used for the two courses on 'Fundamentals of steelmaking metallurgy' offered at Hoogovens IJmuiden during 1988–91. We acknowledge the help in the editing and text corrections from Ing A. B. Snoeijer, Ing W. van der Knoop, Ir J. Van der Stel, Ir A. van der Heiden, Ir W. Tiekink, and Drs G. Abbel, all belonging to the Research and Works Laboratories at Hoogovens IJmuiden. We are extremely grateful to Dr R. L. Moss of the Commission of the European Communities, Joint Research Centre, Petten Establishment, Petten, The Netherlands, for going through the first three chapters and suggesting several corrections. We sincerely thank Professors M. L. Vaidya, R. Shekhar and D. Mazumdar, all at The Department of Metallurgy, IIT, Kanpur, for corrections to and useful suggestions for Chapters 4–8.

We thank Mr J. K. Mishra and Mr U. S. Misra of IIT, Kanpur, for their assistance in producing a major part of the typescript. We also acknowledge help from Mrs P. Iskes-de Jong, Miss I. Wesseling and Miss M. Langereis, all at Hoogovens IJmuiden, in typing the tables, figure captions and references.

We also thank Mr W. A. M. de Boer of Hoogovens IJmuiden and Mr R. K. Bajpai of IIT, Kanpur, for their excellent work in tracing almost all the line drawings which appear in this book.

Finally, our very special thanks are due to our families, especially to Mrs Ellen Boom-Vrijhoef and to Mrs Surabhi Shukla, for their moral support and patience with our long hours spent working on the book.

B.D. and R.B.

List of symbols

A	area, m^2; nominal interfacial area; cross-sectional area
A_D	fractional contribution of diffusion mass transfer
A_e	area of the nozzle at exit
A_{op}	cross-sectional area of plume
A_t	area of the nozzle at throat
A_{tot}	sum of nozzle area, m^2
ATOXi	molecular weight of oxide
ATSUi	molecular weight of sulphide
a, a_R	Raoultian activity
B	basicity
b	(sub- or superscript) bulk
C	number of components in a system; constant; concentration, mol/m^3
$C-$	concentration, i.e. C_A = concentration of A, C_B = concentration of B
C_g^b	concentration in the bulk in gas
C_g^i	concentration at the interface in gas
C_i	number of circulations made by a fluid element
C_m^b	concentration in the bulk in metal
C_m^{eq}	concentration in metal at equilibrium
C_m^i	concentration at the interface in metal
C_p	heat capacity (per mole at constant pressure)
C_{P_M}	phosphorus capacity
$C_{PO_4^{3-}}$	phosphate capacity
C_S	sulphide capacity
C_s^b	concentration in the bulk in slag
C_s^{eq}	concentration in slag at equilibrium
C_s^i	concentration at the interface in slag
C_v	heat capacity (per mole) at constant volume
C_{Vm}	molar volume of metal
C_{Vs}	molar volume of slag
C_1	empirical constant
C_2	empirical constant

C_μ	universal energy dissipation rate constant
c	nozzle constant
D	total derivative
D	vessel/bath diameter; diffusion coefficient
$D_{CO_2/CO}$	binary diffusivity of CO_2 and CO
D_{eff}	effective diffusion coefficient (fluid in motion)
D_m	molecular (or atomic) diffusion coefficient
D_t	eddy or turbulent diffusion coefficient
d	diameter, m; inner diameter of orifice
d'	measure of the fineness of the particles
d_b	bubble diameter
$d_{b\text{-crit}}$	critical diameter of the bubble before disintegration
d_e	exit diameter, m
d_{limit}	maximum diameter of the droplet considered in the drop size distribution
d_o	outer nozzle diameter of non-wetted submerged nozzle or throat diameter of de Laval (supersonic) nozzle
d_t	throat diameter, m
d_{vs}	volume surface mean diameter of bubbles
E	activation energy
E_i	Energy of state
Eo	Eötvös number
e_j^i	Wagner's interaction parameter, wt % basis
F	flow rate, m^3/min; degrees of freedom of a system
F	body force
Fl	fluidity constant
Fr	Froude number
Fr'	modified Froude number
FMETAL	weight of metal at equilibrium
FSLAG	weight of slag at equilibrium
f	fugacity; fraction of particles which reside inside the bubble; Henrian activity coefficient
f	(subscript) signifies permanent reaction at slag–metal interface
f_i	Henrian activity coefficient of solute i
G_k	rate of production of k
G^o	free energy in standard state
g	acceleration due to gravity, m/s^2
H	enthalpy; bath height or specific enthalpy of gas
H^{absorb}	heat of absorption
h	foam height, m; plume height or the distance of the nozzle from bath surface
h_d	projected plume section above bath surface, m
h_{dl}	dimensionless lance height
h_i	Henrian activity of solute i, wt % basis

h'_i	Henrian activity of solute i, mole fraction basis
I_x	specific blowing impact at distance x from nozzle tip
i	(sub- or superscript) interface
J_D	total flux in metal
J_{D_m}	diffusion flux in metal
J_{D_t}	turbulent flux in metal
J_m	flux density of solutes in metal (except oxygen and sulphur)
J_O	flux density of oxygen in metal
J_S	flux density of sulphur in metal
J_s	total flux in slag
j	flux, mol/m² s
K	constant; equilibrium constant, i.e. KOX_i = equilibrium constant for oxide, KSU_i = equilibrium constant for sulphide; 'i' refers to the particular component involved
K'	equilibrium quotient
K_e	absorption equilibrium constant
$\mathcal{K}$	capacity coefficient of mass transfer
$\mathcal{K}_{me}$	mass transfer capacity index in metal, m/s
$\mathcal{K}'_{me}$	volumetric mass transfer coefficient in metal (m³/s)
$\mathcal{K}_{sl}$	mass transfer capacity index in slag, m/s
$\mathcal{K}'_{sl}$	volumetric mass transfer coefficient in slag (m³/s)
k	rate constant; mass transfer coefficient, m/s (subscript Al = aluminium in metal, C = carbon in metal, O = oxygen in metal, b = bubbles, f = reaction at slag–metal interface, g = gas, m = metal, p = in metal around particles, sl = slag); turbulent kinetic energy of fluctuating motion
k_b	backward reaction rate constant
k_f	forward reaction rate constant
k_g	overall mass transfer coefficient in gas phase
k_g^D	diffusion mass transfer coefficient in gas
k_m	overall mass transfer coefficient in metal phase
k_m^D	diffusion mass transfer coefficient in metal
k_m^E	turbulent mass transfer coefficient in metal
k_{me}	mass transfer index in metal, m/s
k_{mM}	mass transfer coefficient of solutes in metal (except oxygen and sulphur)
k_{mO}	mass transfer coefficient of oxygen in metal
k_{mS}	mass transfer coefficient of sulphur in metal
k_{OX}	mass transfer coefficient of oxides in slag
k_r	forward reaction rate constant, mol/m² s atm; reaction rate constant
k_s	apparent equilibrium constant; overall mass transfer coefficient in slag phase
k_{sl}	mass transfer index in slag, m/s
k_{SU}	mass transfer coefficient of sulphides in slag
L	partition coefficient; characteristic length
L_p	partition or distribution coefficient

M	atomic weight
Ma	Mach number
M–	moles of solute at equilibrium, i.e. MSU = moles of sulphur, MNT = moles of nitrogen
M_e	Mach number at nozzle exit
M_g	molecular weight of gas; iron conversion, tons/min
M_i	atomic weight of solute i
MO_2	molecular weight of oxygen
$\dot{m}$	mass flow rate of gas
$\overset{*}{\dot{m}}_t$	maximum mass flow rate of gas
N	ionic fraction
N'	electrically equivalent ion fraction
n	number of moles; number of nozzles; measure of the homogeneity of particle size distribution
n_r	number of restrictions imposed on a system
O_b	mole fraction of oxygen in bulk
O_i	mole fraction of oxygen at interface in metal
OX_b	mole fraction of oxide in bulk
OX_i	mole fraction of oxide at interface
p	pressure, atm; dimensionless product
P	number of phases in a system
Pe_m	Peclet diffusion number
Pe_t	Peclet turbulent number
$p_{d,e}$	dynamic pressure at nozzle exit
$p_{d,x}$	dynamic pressure at distance x from nozzle tip
p_e	pressure of gas at nozzle exit
p_o	inlet pressure, atm; pressure of gas in the pipe
p_t	pressure of gas at the throat of the nozzle
Q	volumetric flow rate of gas, m^3/s
Q_{bath}	flow rate at temperature of bath, m^3/s
$\dot{q}$	rate of heat generation
q_e	kinetic energy of gas at nozzle exit
$q_{o,x}$	kinetic energy of gas at distance x from nozzle exit
R	gas constant
Re	Reynolds number
Re_b	bubble Reynolds number
R_g	recirculation ratio
R_{ij}	fraction of cells of type ij
R_v	vessel diameter
RF	drop size distribution function
RRS	Rosin–Rammler–Sperling distribution function
r	radial coordinate; radius of jet, m
r_o	overall rate; jet radius at which gas velocity in the jet is half the axial velocity
r_s	rate through covered sites

r_t	radius of the throat of the nozzle
r_u	rate through bare sites
$r_{\alpha max/2}$	half value radius
S_b	mole fraction of sulphur in bulk metal
S_f	surface renewal factor
S_i	mole fraction of sulphur at interface in metal
S^o	standard entropy
Sc	Schmidt number
Sh	Sherwood number
SU_b	mole fraction of sulphide in bulk slag
SU_i	mole fraction of sulphide at interface in slag
s	velocity of sound
T	temperature, K
T_{bath}	bath temperature, K
T_e	temperature of gas at exit
T_g	temperature of gas film, K
T_l	liquid temperature
T_o	temperature of gas in pipe
T_t	temperature of gas at throat
TIM	turbulence index of metal
TIS	turbulence index of slag
TMETAL	total number of moles of metal at equilibrium
TSLAG	total number of moles of slag at equilibrium
t	time, s
t_c	circulation time
trb	time of rise of bubbles
trp	time of rise of particle
U_s	superficial velocity of gas, cm/s
u	velocity, u_r and u_z indicate the components in the radial and z direction, respectively
$\mathbf{u}$	velocity vector
u_b	rising velocity of the bubble
u_d	exit or free-space velocity of gas at orifice (same as U_s except for the units)
u_n	velocity of liquid normal to interface
u_o	axial velocity
$u_{o,x}$	centreline velocity of gas at distance x from the nozzle tip
$u_{r,x}$	velocity of gas at radius r and distance x from the nozzle tip
V	volume, m^3; volume of metal, m^3; molar volume; index for vapour
V_{Ar}	volumetric flow rate of argon, $m^3 s$
V_{H_2}	volumetric flow rate of hydrogen, $m^3 s$
V_m	volume of metal
V_{met}	mixing rate of metal, tons/min
V_s	volume of slag

V_{slag}	mixing rate of slag, tons/min
$\dot{V}_g$	gas flow rate, cm^3/s
W	mass of gas; wt % of solute, i.e. WMn $=$ wt % of Mn, WSi $=$ wt % of Si
W_1	kinematic flow rate of buoyancy
We	Weber number
WK	work done
X	mole fraction; i.e. XC $=$ mole fraction of carbon, XS $=$ mole fraction of sulphur; $X_i =$ etc. mole fraction of component i, where 'i' stands for Fe, Mn, Si, Al, etc.
x	distance of metal bath from lance tip, m
x_{ss}	length of the supersonic core

Greek symbols

α	gas volume fraction in two-phase gas–liquid region
α_{max}	gas volume fraction at centreline
γ	gamma phase in iron; ratio of the specific heats of a gas; Raoultian activity coefficient
γ_i	Raoultian activity coefficient of solute i
γ_i°	Raoultian activity coefficient at infinite dilution
δ	boundary layer thickness
δ_{ij}	Kronecker delta symbol
ε	coefficient of flow rate; rate of energy dissipation
$\dot{\varepsilon}$	mixing power density; stirring energy input rate, W/ton
$\dot{\varepsilon}_{bs}$	energy dissipation in bubble slip
$\dot{\varepsilon}_c$	energy dissipation in liquid recirculation
ε_i^j	interaction parameter, mole fraction basis
ε_1	dimensionless density
ε_o	dimensionless density difference
ζ	friction factor
θ	angle of the divergent part of the nozzle; fraction of covered sites by the adsorbed atoms (or molecules)
λ	geometrical scaling factor
μ	viscosity, kg/m s or Pa/s
μ_e, μ_{eff}	effective viscosity
μ_1	viscosity of liquid
μ_m	molecular viscosity
μ_t	eddy viscosity; turbulent viscosity
v	kinematic viscosity
ρ	density, kg/m^3
ρ_e	density of gas at nozzle exit
ρ_g	density of gas

ρ_{go}	density of two-phase gas–liquid mixture
ρ_l	density of liquid
ρ_m	density of metal
ρ_o	density of the gas in the pipe
ρ_s	density of slag
ρ_t	density of the gas at throat of the nozzle
σ	surface tension, N/m or kg/s^2
σ_c	turbulent Schmidt number
σ_k	effective Prandtl number for k
σ_t	turbulent Prandtl number
σ_ε	effective Prandtl number for ε
ϕ	volume flow rate, m^3/s; dimensionless gas hold-up
ΔC_p	change in heat capacity at constant pressure
ΔC_p^{syst}	change in heat capacity of a system
ΔG	free energy change
ΔG^M	integral molar free energy of mixing
$\Delta \bar{G}_i^M$	relative partial molar free energy of solute i
ΔG^o	free energy change in standard state
$\Delta G^{o,\,overall}$	standard free energy change due to all reactions
$\Delta G^{o,\,react}$	standard free energy of reaction
ΔH	enthalpy change
ΔH^{fuse}	enthalpy of fusion
$\Delta H^{o,\,react}$	standard enthalpy of reaction
ΔH^{syst}	enthalpy change in a system
ΔH^{trans}	enthalpy of transformation
$\Delta \bar{H}_i^M$	relative partial molar enthalpy of solute i
ΔS^{fuse}	entropy of fusion
$\Delta S^{o,\,react}$	entropy change in a reaction
$\Delta \bar{S}_i$	relative partial molar entropy of solute i
ΔV	change in atomic or molar volume on mixing
$\Delta \bar{V}_i^M$	relative partial molar volume of solute i
Λ	optical basicity
Ω	configurational partition function

Other symbols

[ ]	solute in metal phase
( )	slag
( )$_g$	gas
{ }	liquid
$\langle\,\rangle$	solid

Physical fundamentals

1.1 Phases and interfaces encountered in steelmaking processes

The fascinating aspect of steelmaking is that it deals with the dynamic interaction among ions, atoms and molecules which may exist simultaneously at the interfaces of the immiscible slag, metal and gas phases at high temperature. Liquid slag consists of a solution of ions or complex molecules; gas consists of free atoms or molecules; liquid metal consists of a solution of atoms. Since metal, slag and gas are immiscible in each other, the three of them can meet simultaneously only across a line, while any two of them can meet across a plane or curved surface. The number of interfaces which can exist between two given phases at a time can be numerous and variable.

In a stationary situation liquid slag, which is lighter than metal, floats on top of liquid metal whereas in an agitated bath the metal droplets may become entrained in the liquid slag. Alternatively, under the influence of intense turbulent mixing caused by gas stirring or powder injection, the slag droplets or solid slag particles may be drawn into the liquid metal. The same applies to the coexistence of metal–gas and slag–gas phases. The gas can be present as an atmosphere above the slag and liquid metal, or as bubbles entrained in the metal or slag, or as a gas plume rising through the liquid metal and slag, or as a jet impinging upon the metal and slag, etc. As an illustration, Figure 1.1 shows a typical slag–metal–gas system in steelmaking. The numbers inside the circles mark the six possible reaction interfaces between metal, slag and gas, as shown below.

Type of interface	Encircled region in Figure 1.1
Slag–metal interfaces	1, 7
Metal–gas interfaces	3, 6, 8, 9
Slag–gas interfaces	2, 4, 10
Gas–refractory interfaces	5, 11, 12
Metal–refractory interface	13
Slag–refractory interface	14

1

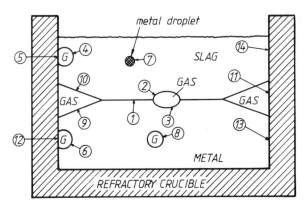

Figure 1.1   Phases and interfaces in slag–metal–gas reactions; G = gas. For the explanation of different numbers see text.

1.2   Physical properties of metal and slag

Liquid iron usually contains several dissolved elements like carbon, silicon, manganese, phosphorus, sulphur, hydrogen, oxygen and nitrogen. The slag phase, liquid or solid (including the refractories), may consist of the oxides, carbides, sulphides, fluorides and chlorides. The gas phase may contain oxygen, hydrogen, carbon dioxide, carbon monoxide, nitrogen and argon.

The physical properties of liquid iron, like viscosity, density and surface tension, change gradually with temperature (Table 1.1) [1–3]. The density (ρ) of pure liquid iron, up to approximately 1973 K, can be described [2] by the equation

$$\rho = (8.537 - 0.001\,15T) \times 10^3\,\text{kg/m}^3 \tag{1.1}$$

where T is the temperature in kelvin.

The viscosity of the binary liquid Fe–C, Fe–Si and Fe–Mn alloys has been determined by many investigators [4, 5]. The average values in the temperature range 1823–1873 K are summarized in Table 1.2. Data on the molecular diffusivity of the elements in binary and multicomponent liquid iron alloys are readily available [2].

Table 1.1   Viscosity, density and surface tension of pure liquid iron at different temperatures.

Pure iron	1823 K	1873 K	1923 K	Ref.
Viscosity (kg/m s $\times 10^3$)	5.5	5.1	4.7	[2]
Density (kg/m^3 $\times 10^{-3}$)	7.07	7.01	6.95	[1]
Surface tension (J/m^2 $\times 10^3$)	1800 ± 20	1790 ± 20	1780 ± 20	[3]

Table 1.2 Viscosity ($kg/m\,s \times 10^3$) of binary alloys at 1823–1873 K.

Dissolved element	wt % of dissolved element				Ref.
	1 wt %	2 wt %	3 wt %	4 wt %	
Carbon	4.5	4.0	3.7	3.3	[4]
Silicon	4.80	4.70	4.65	4.60	[4]
Manganese	7.0	6.8	6.6	6.4	[5]

The steelmaking slags are complex because they may contain several oxides (CaO, MgO, Na_2O, Fe_2O_3, FeO, MnO, SiO_2, Al_2O_3, P_2O_5 and TiO_2), sulphides (CaS, FeS and MnS), fluorides (like CaF_2), chlorides (like $CaCl_2$ and NaCl) and carbides (CaC_2). The attraction or repulsion between the different components depends strongly on temperature and on the differences between the electronegativities of the cations and anions. For example, the bond between the sodium and chlorine ions in pure NaCl is nearly 100% ionic. In pure CaF_2 the bond between the calcium and fluorine ions is approximately 78% ionic and 22% covalent; in pure CaO the ratio of ionic to covalent bonding is 60:40; whereas in SiO_2 and P_2O_5 the ratio is 36:64 and 28:72, respectively. The two approaches, namely the molecular theory of slag and the ionic theory of slag, are commonly employed (see Chapter 2) to study the thermodynamics of complex liquid slags.

The phase diagrams, physical properties like viscosity, density, surface tension, electrical conductivity and interfacial tension, as well as the thermal properties of the steelmaking slags have been extensively reviewed [6–10]. The role played by the physical properties of the slags in the thermodynamics and kinetics of refining reactions during steelmaking will be discussed in the following chapters.

1.3 Physical interaction between gas and liquid metal

The chemically reactive gases like hydrogen, nitrogen and oxygen can dissolve in liquid iron in the atomic state; the atomic state is represented as [H], [N] and [O], respectively. When the solubility limits of hydrogen and nitrogen are exceeded, bubbles of molecular hydrogen and nitrogen (gas phase) can form and grow at suitable nucleation sites. After attaining a certain critical size the bubbles detach themselves from the nucleation sites and try to float to the surface as a result of the buoyancy forces acting on them. If the liquid metal is in the process of solidification the bubbles can be trapped as blow holes. Dissolved nitrogen can also react with dissolved alloying elements in steel; for example, dissolved aluminum [Al] and titanium [Ti] can react with dissolved nitrogen [N] to form solid nitrides.

Oxygen gas has a stronger chemical affinity for iron than hydrogen and nitrogen. As soon as the solubility limit of oxygen is exceeded it can form iron oxide even when iron is in the liquid state. The iron oxide, being lighter than iron, floats to the top as a separate slag phase. If the liquid iron contains other alloying elements like carbon, silicon and aluminium, then the dissolved oxygen or FeO can react chemically to form carbon monoxide gas, SiO_2 and Al_2O_3 (see Chapters 2–7). The carbon monoxide gas, similar to hydrogen and nitrogen, can be trapped as blow holes in a solidifying liquid.

The bubbles, given freedom, rise vertically upwards by displacing the liquid iron from the top to the edges of the bubble. The liquid iron is drawn from the lower regions as well as from the sides to fill up the void left behind by the rising bubbles (bubble wakes). The stirring action, thus set up due to rising gas bubbles, enhances mass transfer, fluid flow and heat transfer. If the liquid iron is covered with a layer of slag, then at the interface of the slag and metal the rising bubble may carry a thin film of metal up to some distance in the slag. The film soon breaks and the liquid metal consolidates as droplets. Since the metal droplets are heavier than slag they fall down through the slag layer. The slag–metal interfacial area for chemical reactions increases as does the interface mass transfer. As the bubble frequency crossing the slag–metal interface increases the relative amount of metal film carried by the bubble decreases. The intermixing of slag and metal phases, however, increases.

Despite a strong theoretical base available to describe mathematically the physical interaction between gas and metal from first principles [11–16], we are often forced to use empirical correlations. This is primarily due to the difficulties faced in carrying out accurate experiments at the steelmaking temperatures (1500–1900 K), as well as to the stochastic and dynamic nature of the steelmaking processes. The physical phenomena associated with the formation and rise of gas bubbles during the submerged injection of gases like argon and nitrogen are discussed in this chapter. The thermodynamics and kinetics of gas–metal and gas–slag reactions in steelmaking will be discussed in Chapters 2–8.

1.3.1   *Bubble growth, detachment and rise*

At small gas flow rates, the bubbles form at the tip of the submerged nozzle (orifice) and grow in size as long as the buoyancy force acting on the bubble is smaller than the force of surface tension:

$$\tfrac{1}{6}\pi d_b^3 g(\rho_1 - \rho_g) < \pi d_o \sigma \tag{1.2}$$

where d_o is the outer diameter of the non-wetted nozzle (in the case of wetted nozzles the inner diameter is used because the bubble then grows at the inner periphery), σ is the surface tension, ρ_1 and ρ_g are the densities of liquid and gas respectively, and d_b is bubble diameter. The bubble detaches from the orifice when the buoyancy force exceeds the surface tension. From Equation (1.2) the limiting diameter of a spherical bubble in a non-wetting system is given by

$$d_b = \left(\frac{6 d_o \sigma}{g(\rho_1 - \rho_g)} \right)^{1/3} \tag{1.3}$$

The rising velocity of gas bubbles depends on the balance between the buoyancy force and the total drag (viscous force and form drag). When these forces are balanced the bubbles rise at a constant velocity, called the terminal velocity. The dimensionless numbers used to characterize the motion of the bubbles are the bubble Reynolds number (Re_b), the Weber number (We) and the Eötvös number (Eo), where

$$Re_b = \frac{d_b u_b \rho_1}{\mu_1} \tag{1.4}$$

(Equation (1.4) is true for $Re_b < 400$.)

$$We = \frac{d_b u_b^2 \rho_1}{\sigma} \tag{1.5}$$

$$Eo = \frac{d_b^2 g(\rho_1 - \rho_g)}{\sigma} \tag{1.6}$$

At the tip of the orifice, where the bubbles form, the Weber number can be written as

$$We = \frac{d_o u_d^2 \rho_1}{\sigma} \tag{1.7}$$

where d_o is the outer diameter of the orifice (non-wetted) and u_d is the exit velocity of the gas at the orifice (also called superficial velocity). The Froude number (Fr) at the orifice is defined as

$$Fr = \frac{u_d^2}{g d_o} \tag{1.8}$$

For gas–liquid systems, a modified Froude number (Fr') is also employed:

$$Fr' = \left(\frac{u_d^2}{gd}\right)\left(\frac{\rho_g}{\rho_1 - \rho_g}\right) \tag{1.9}$$

The rising velocity and geometrical shape of the bubbles depend upon the diameter of the bubbles [16]. Since there is almost no internal circulation of gas due to surface tension forces in bubbles of diameter $\leqslant 0.2$ cm and $Re_b \leqslant 1$, they behave like rigid spheres. Such small bubbles rise vertically at a terminal velocity (u_t) determined by Stokes' law:

$$u_t = g\left(\frac{d_b^2}{18\mu_1}\right)(\rho_1 - \rho_g) \tag{1.10}$$

Intermediate-size spherical bubbles (0.2 cm $< d_b < 0.5$ cm, $1 < Re_b < 400$) rise at a greater velocity than predicted by Stokes' law. The drag force on the bubble is reduced owing to the circulation of gas within the bubble and the displacement of liquid at the bubble–metal interface. With increasing size and velocity (0.5 cm $< d_b < 1$ cm, $400 < Re_b < 1000$ and $We > 3.2$) the combination of viscous, buoyancy and drag forces acting on the bubble may gradually change the bubble's shape from spherical to spheroidal and ellipsoidal (d_b in such cases is the equivalent diameter). Rising non-spherical bubbles move upwards, but in a spiral path. Bubbles of equivalent diameter greater than 1 cm ($Re_b > 1000$, $We > 18$ or $Eo > 50$) take the shape of a spherical cap and rise at a velocity almost independent of the properties of the fluid.

Several laboratory and pilot-scale experiments have been reported to determine the empirical correlations for the size and the rising velocity of gas bubbles in liquids. The correlation obtained for one liquid may not be applicable to another liquid because of differences in chemical composition and physical properties. For example, the bubble size in non-deoxidized and non-desulphurized steels is smaller than in killed and desulphurized steels; smaller bubbles are produced in liquid iron than in slag. It has been suggested that for the results of the cold or hot physical modelling experiments to be applicable to industrial iron melts, the products of *Re*, *We* and *Fr* expressed as

$$Re^4 Fr We^{-3} = Fl = \rho_1 \sigma^3 g^{-1} \mu^{-4} \tag{1.11}$$

should be of the same order of magnitude [17]. *Fl* is called the fluidity constant and is the inverse of the Morton number. The value of *Fl* for physical modelling (cold) experiments involving water at 293 K is 3.9×10^{10}.

For liquid steel at 1873 K, typical values to be substituted in Equation (1.11) are $\rho_1 = 7.02 \times 10^3$ kg/m^3, $\mu = 5.0 \times 10^{-3}$ kg/m s and $\sigma = 1.2$ kg/s^2. The value of *Fl* is thus 1.98×10^{12}.

It is obvious that the value of *Fl* for liquid steel is approximately 50 times greater than that for water. As a result of this disparity, the results obtained on water models are not directly applicable to liquid steel. In most cases, adapted empirical equations are used for the real processes.

Over a wide range of gas flow rates and bubble sizes (in the range of 0.07 to 4 cm diameter), the following empirical equation, as reported in the original paper in c.g.s. units, can be used to estimate the bubble diameter in liquid iron [18]:

$$d_b = \left[\left(\frac{6\sigma d_o}{\rho_1 g} \right)^2 + 0.0242 (\dot{V}_g^2 d_o)^{0.867} \right]^{1/6} \tag{1.12}$$

where d_b is the bubble diameter (cm), d_o is the nozzle diameter (cm), $\dot{V}_g$ is gas flow rate (cm^3/s), σ is surface tension (g/s^2), g is acceleration due to gravity (cm/s^2) and ρ_1 is the density of liquid (g/cm^3).

The rising velocity of a bubble in liquid iron can be calculated from [19]

$$u_b = \left(\frac{2\sigma}{\rho_1 d_b} + \tfrac{1}{2} g d_b \right)^{0.5} \tag{1.13}$$

where u_b is the rising velocity of the bubble (cm/s), σ is surface tension (g/s^2), d_b is the bubble diameter (cm), g is acceleration due to gravity (cm/s^2) and ρ_1 is the density of liquid (g/cm^3).

Example 1.1 Diameter and velocity of bubbles rising in liquid iron

Calculate the diameter and the rising velocity of a bubble in liquid iron, given the following:

- diameter of nozzle, $d_o = 0.505$ cm
- volumetric flow rate of gas through the nozzle, $\dot{V}_g = 2083$ cm^3/s
- surface tension $\sigma = 1540$ g/s^2
- acceleration, $g = 981$ cm/s^2
- density, $\rho = 7$ g/cm^3

On substituting the requisite values in Equation (1.12), the bubble diameter can be calculated as

$$d_b = \left[\left(\frac{6 \times 1540 \times 0.505}{7 \times 981} \right)^2 + 0.0242(2083^2 \times 0.505)^{0.867} \right]^{1/6}$$

$$= 4.43 \text{ cm (or } 0.0443 \text{ m)}$$

On substituting the requisite values in Equation (1.13), the rising velocity can be calculated as

$$u_b = \left(\frac{2 \times 1540}{7 \times 4.43} + \tfrac{1}{2} \times 981 \times 4.43 \right)^{0.5}$$

$$= 47.14 \text{ cm/s (or } 0.47 \text{ m/s)}$$

The bubbles rising in a liquid increase in size due to the lowering of the static pressure head of liquid. An increase in size leads to a greater surface tension acting on the bubble. When the force of inertia of the upward-moving bubble exceeds the surface tension, then necking occurs and the bubble disintegrates into smaller sizes. The critical diameter, $d_{\text{b-crit}}$, for the disintegration of spherical cap bubbles can be estimated [20] from

$$d_{\text{b-crit}} = \left[2 \left(\frac{6}{\xi} \right)^{1/3} \frac{\sigma}{0.52 g (\rho_g \rho_l^2)^{1/3}} \right]^{1/2} \tag{1.14}$$

where ξ is the friction factor [21] (Figure 1.2).

Example 1.2 Critical diameter for the disintegration of spherical cap bubbles of carbon monoxide gas rising in liquid steel

Given the following:

- the density of carbon monoxide gas at 1873 K, $\rho_g = 0.182$ kg/m^3
- at $Re = 1000$, the friction factor (from Figure 1.2) $= 0.44$
- surface tension of steel, $\sigma = 1.2$ kg/s^2
- $Re > 1000$

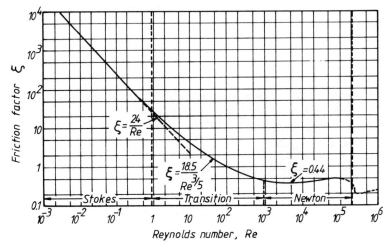

Figure 1.2   Friction factor for solid particles as a function of Reynolds number. (*Source*: Lapple, C. E. (1950) In *Chemical Engineers Handbook* (ed. J. H. Perry), 3rd edn, McGraw-Hill: New York.)

On substituting the above values in Equation (1.14) the maximum value of $d_{b\text{-crit}}$, before the disintegration of carbon monoxide bubbles in liquid steel at 1873 K, is found as 7.34 cm. In a turbulent bath, however, the stability of the bubble is further diminished due to collision with eddies [22, 23] and the critical bubble size for the disintegration to occur can be even lower than predicted by Equation (1.14).

1.3.2   *Bubbling versus jetting and formation of a plume*

The phenomena occurring at the tip of the submerged orifice (or nozzle or tuyère) are of practical significance as far as erosion of the tip of the orifice, wear of the refractory material around the tip and the mixing effect caused by the rising bubbles are concerned (the mixing effect is discussed separately in section 1.4).

At subsonic gas flow rates one bubble at a time is released at the tip (Figure 1.3a). In between the release of two bubbles, the liquid can enter the tuyère; this is termed tuyère flooding. At high temperatures in steelmaking, flooding can cause tuyère wear, and in extreme cases liquid metal can solidify inside the tuyère; this is called tuyère blockage or clogging. The phenomenon of tuyère wear by the surrounding liquid is also known as back-attack or back-wash. The back-attack frequency increases as the gas flow rate and hence the bubble frequency increase.

As the linear exit velocity of the gas at the tip further increases, the forces of surface tension acting on the bubble forming at the tip are overcome by the force of inertia and the bubble begins to elongate in the vertical direction without detaching

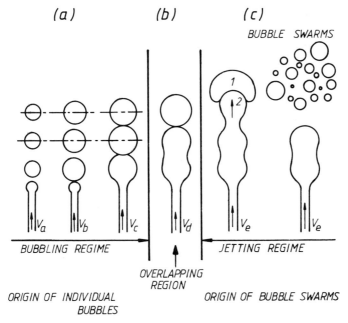

Figure 1.3 Schematic description of bubble and jet formation during gas injection into liquid metal. (*Source*: Steinmetz, E. and Scheller, P. R. (1987) *Stahl und Eisen*, **107**, 417–25.)

itself from the tip (Figure 1.3b). Ultimately, with increasing values of We^2/Fr at the tip, the elongation of the bubble in the vertical direction leads to the formation of a continuous jet. The bubbling/jetting transition occurs over a small range of gas flow rate. At sonic flow the jet is still unstable and collapses frequently. Back-attack, however, becomes significantly smaller as the jet stabilizes at still higher gas flow rates. In the jetting regime the back-attack on rectangular orifices is found to be much less than on circular orifices (or tuyères).

At some distance from the tip of the nozzle, if the bath is deep enough, the jet eventually breaks up into a swarm of rising bubbles. The diffuse cone of liquid which contains the bubbles is called the bubble plume.

In most industrial applications, the jet and bubble plume can form even at low gas flow rates of 0.015 m³ (stp)/min [24], especially when a cluster of small tuyères are used, or if a mushroom of solidified metal forms at the tuyère tip, or if porous bricks are used. In the latter two cases (i.e. mushroom and porous bricks), it is difficult to establish a criterion for the bubbling/jetting transition because the pore diameter is too small.

The size of bubbles contained in a plume differs from those of individual bubbles rising at different locations in the liquid metal; the bubbles in a plume have a greater opportunity to disperse or to coagulate and thus form a range of sizes of bubbles. In the cold model experiments, it is observed that the bubble plume rises vertically upwards in a swirling motion. This results in waves at the top surface. In liquid metals the swirling action is not visible but the wave motion of the fluid on the top surface is observed when the orifices or tuyères are symmetrically placed at the bottom.

The following empirical equation, reported in c.g.s. units in the original paper [25], has been proposed to predict the size of bubbles rising in bubble swarms:

$$d_{vs} = 0.091 \left(\frac{\sigma}{\rho_1}\right)^{0.5} U_s^{0.44} \tag{1.15}$$

where d_{vs} is the volume surface mean diameter (cm) of bubbles, U_s is the superficial velocity of gas (cm/s), ρ_1 is the density of liquid (g/cm^3) and σ is surface tension (g/s^2).

1.4   Mixing phenomena in a gas-stirred molten-metal bath

Small gas bubbles, as mentioned earlier (see Equation (1.10)), behave like rigid spheres and are able to rise in the liquid without causing much disturbance; small bubbles do displace the liquid when they move upwards but the disturbance thus introduced is only local. As the bubble size increases, the rising velocity also increases. Large, irregular-shaped bubbles (namely, spherical cap bubbles) entrain a certain amount of liquid with them as they rise. A gas–liquid plume is thus formed which causes a recirculatory flow and the liquid which moves up with the gas bubble has to come down in a recirculation pattern along the walls of the vessel and around the region of the plume itself. The recirculatory flow results in a thorough mixing of the liquid in the vessel. At higher gas velocities, the gas jet formed at the nozzle tip also breaks down into a two-phase gas mixture – called a buoyant jet or a forced plume. The mixing behaviour of a forced plume is very different from an ordinary plume. Mathematical treatments of fluid flow and mixing phenomena have been reported in the literature, in general, for simplified geometries. A comprehensive review of the mixing theories of gas-stirred melts and the mixing phenomena in metallurgical vessels is provided in Oeters *et al.* [26].

Two different gas injection methods can be employed; the porous plugs or tuyères can be attached to the bottom or side walls of the vessel in symmetric or asymmetric positions; alternatively, a lance fitted with a nozzle (or nozzles) at the tip can be vertically (or at an angle) lowered into the metal bath up to a desired depth. The nozzles can be inclined and placed at suitable locations with respect to the r, θ and z coordinates of the vessel to obtain a maximum effect of the mixing induced by recirculatory flow. In some processes, chemically reactive gases and/or chemically reactive powders or both are co-injected with inert gases to achieve mixing, desulphurization, desiliconization and dephosphorization.

The mixing caused by the recirculatory flow is quantified by the mixing time τ (in seconds) required to achieve a certain degree of concentration homogenization throughout the liquid in the vessel. According to the fundamental definition, the degree of mixing, γ, is given by

$$\gamma = \frac{C - C_i}{C_f - C_i} \qquad (1.16)$$

where C_i is the initial uniform concentration of a species, C_f is the final uniform concentration and C is the concentration corresponding to mixing time τ. If $C_i = 0$, then $\gamma = C/C_f$. In the literature, the time required to achieve 95% homogenization ($\gamma = 0.95$) is usually quoted as mixing time, τ. In a dimensionless form, the mixing effect is also expressed as a circulation number defined as the number of circulations a liquid element makes in the metal before the bath achieves 95% homogenization.

Three separate approaches have been followed to model the mixing phenomena:

1. Turbulence models, based on the solution of the Navier–Stokes equation pertaining to turbulent recirculatory flows.
2. Circulation models, based on a simplified energy balance in physical systems; a good experimental correlation between mixing time and circulation time (determined from the energy balance) has generally been observed.
3. Empirical models, based on dimensional analysis and/or simple empirical fitting of the experimental data to correlations.

All the above approaches employ several simplifying assumptions. In most cases, for ease of experimentation, water or cold models are used. The empirical relations developed with the help of cold model experiments for simple geometries are then adapted to commercial gas-stirred ladles or furnaces containing liquid steel.

1.4.1 Turbulence models

The turbulence models based on the solution of the Navier–Stokes equation provide a good understanding of the mixing process on the basis of fundamental fluid flow equations. It follows from these models that the intensity of mixing is determined by the dissipation rate of turbulent kinetic energy. For a wide range of power inputs in bottom-stirred metallurgical systems [27], the mixing time can be expressed by an empirical relation

$$\tau = 12\,680\dot{\varepsilon}^{-1/3} \qquad (1.17)$$

where $\dot{\varepsilon}$ is the stirring energy input in W/ton.

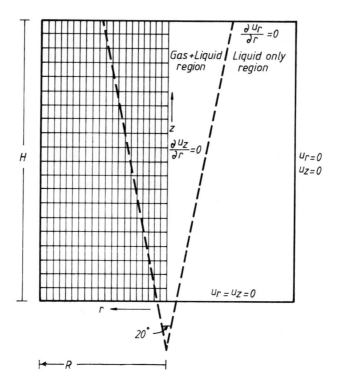

Figure 1.4   Schematic sketch of axisymmetric gas injection through the bottom of a cylindrical vessel; grid arrangement is used in numerical analysis. (*Source*: Woo, J. S., Szekely, J., Castillejos, A. H. and Brimacombe, J. K. (1990) *Metall. Trans.*, **21B**, 269–77.)

Theoretical analysis of the diffusion and dispersion of additions in steel melts [28] has confirmed that at low Peclet numbers, $Pe < 5$ (where $Pe = uL/D_{eff}$, u is characteristic flow velocity, L characteristic length of liquid to mix and D_{eff} effective diffusion coefficient), mixing occurs predominantly by turbulent diffusion; at high Peclet numbers, $Pe > 20$, the contribution of turbulent convection increases and mixing times decrease.

If axisymetric flow is assumed (Figure 1.4), the governing equations for turbulent flow (at steady state and in cylindrical coordinates assuming axial symmetry) can be written as follows.

Equation of continuity

$$\frac{1}{r}\frac{\partial}{\partial r}(\rho r u_r) + \frac{\partial}{\partial z}(\rho u_z) = 0 \tag{1.18}$$

Equation of motion in the axial direction

$$\frac{\partial}{\partial r}(\rho r u_r u_z) + \frac{\partial}{\partial z}(\rho u_z^2) = -\frac{\partial p}{\partial z} + \frac{1}{r}\frac{\partial}{\partial r}\left(r\mu_{\text{eff}}\frac{\partial u_z}{\partial r}\right)$$

$$+2\frac{\partial}{\partial z}\left(\mu_{\text{eff}}\frac{\partial u_z}{\partial z}\right) + \frac{1}{r}\frac{\partial}{\partial r}\left(r\mu_{\text{eff}}\frac{\partial u_r}{\partial r}\right)$$

$$+\rho g\alpha \tag{1.19}$$

where α is the gas volume fraction in the two-phase region, μ_{eff} is effective viscosity, p is pressure, u is velocity, r is radius and g is acceleration due to gravity.

Equation of motion in the radial direction

$$\frac{1}{r}\frac{\partial}{\partial r}(\rho r u_r^2) + \frac{\partial}{\partial z}(\rho u_r u_z) = -\frac{\partial p}{\partial r} + \frac{2}{r}\frac{\partial}{\partial r}\left(r\mu_{\text{eff}}\frac{\partial u_r}{\partial r}\right)$$

$$+\frac{\partial}{\partial z}\left(\mu_{\text{eff}}\frac{\partial u_r}{\partial z}\right) - \frac{\partial}{\partial z}\left(\mu_{\text{eff}}\frac{\partial u_z}{\partial z}\right)$$

$$-\frac{2u_r\mu_{\text{eff}}}{r^2} \tag{1.20}$$

Equation for turbulent kinetic energy *k*

$$\frac{1}{r}\frac{\partial}{\partial r}(\rho u_r k) + \frac{\partial}{\partial z}(\rho u_z k) = \frac{1}{r}\frac{\partial}{\partial r}\left(r\frac{\mu_{\text{eff}}}{\sigma_k}\frac{\partial k}{\partial r}\right)$$

$$+\frac{\partial}{\partial z}\left(\frac{\mu_{\text{eff}}}{\sigma_k}\frac{\partial k}{\partial z}\right) + G - \rho\varepsilon \tag{1.21}$$

where σ_k is the effective Prandtl number for k and G is the generation of turbulent kinetic energy which can be expressed as

$$G = \mu_t\left\{2\left[\left(\frac{\partial u_r}{\partial r}\right)^2 + \left(\frac{\partial u_z}{\partial z}\right)^2 + \left(\frac{u_r}{r}\right)^2\right] + \left(\frac{\partial u_r}{\partial r} + \frac{\partial u_z}{\partial z}\right)^2\right\} \tag{1.22}$$

Equation for dissipation rate of turbulent kinetic energy ε

$$\frac{1}{r}\frac{\partial}{\partial r}(\rho r u_r \varepsilon) + \frac{\partial}{\partial z}(\rho u_z \varepsilon) = \frac{1}{r}\frac{\partial}{\partial r}\left(r\frac{\mu_{\text{eff}}}{\sigma_\varepsilon}\frac{\partial \varepsilon}{\partial r}\right)$$

$$+\frac{\partial}{\partial z}\left(\frac{\mu_{\text{eff}}}{\sigma_k}\frac{\partial \varepsilon}{\partial z}\right) + \frac{C_1\varepsilon G}{k} - \frac{C_2\rho\varepsilon}{k} \tag{1.23}$$

where σ_ε is the effective Prandtl number for ε and C_1 and C_2 are constants.

Boundary conditions

The set of equations (1.18)–(1.23) has to be solved with the following boundary conditions [29] (see Figure 1.4).

Along the axis of symmetry:

$$\frac{\partial u_r}{\partial r} = 0, \quad \frac{\partial k}{\partial r} = 0, \quad \frac{\partial \varepsilon}{\partial r} = 0, \quad u_r = 0 \qquad (1.24)$$

At the free surface:

$$\frac{\partial u_r}{\partial z} = 0, \quad \frac{\partial k}{\partial z} = 0, \quad \frac{\partial \varepsilon}{\partial z} = 0, \quad u_z = 0 \qquad (1.25)$$

At the walls:

$$u_r = u_z = 0 \qquad (1.26)$$

The numerical or analytical solution of the Navier–Stokes equation with imposed boundary conditions has the advantage that we can predict the velocity profiles, without the need to carry out experiments. Experts in numerical analysis have developed software packages like PHOENICS, FLUENT, TEACH(2D, 3D), 2E/FIX, SOLA/VOF, FIDA, and GENMIX with the help of which we can, in principle, solve one-, two- or three-dimensional fluid flow problems. A good level of computational experience and expertise is, however, needed to handle and execute these software packages. In order to get realistic results the imposed boundary conditions must conform as closely as possible to the actual process. In this respect, the real phenomena in primary steelmaking are somewhat elusive because of their turbulent, stochastic and dynamic nature. The chemical composition, interfacial area and physical properties (volume, temperature and mass and diffusion rate) of the reacting metal, slag and gas all change as a function of time. In addition, they follow a unique path each time. The boundary conditions thus constantly change.

It has been found equally rewarding to do cold and hot model experiments in which we can observe, measure and characterize the physical interactions between different phases at different stages of the process as closely as possible [13–16].

Besides the numerical capabilities of the software packages themselves, the accuracy of the predictions made depends upon the choice of viscosity models employed to estimate μ_{eff} and also the simplifying assumptions made to account for the buoyancy in the plume region, as described below.

Viscosity models
The differential model of turbulence (k–ε model) can be used to estimate μ_t. In some cases a combination of the average effective viscosity model as well as the k–ε model is employed [30–33]. The following equation was first used to calculate [30, 31] the bulk effective viscosity, μ_e (kg/m s):

$$\mu_e = KD^{1/2}H^{-1/3}\rho_l^{2/3}(\dot{m}u_d^2)^{1/3} \qquad (1.27)$$

where K is a constant, H is depth of liquid metal (m), ρ_l is the liquid density (kg/m³), D is the vessel diameter (m), $\dot{m}$ is the mass flow rate of gas (kg/s) and u_d is the free-space velocity of gas through the nozzle or the orifice (m/s). In view of the difficulties involved in using Equation (1.27) for gas-stirred systems, the following equation, based on dimensional arguments, was proposed [34]:

$$\mu_e = C\rho_l H\,[Q(1-\alpha)g/D]^{1/3} \tag{1.28}$$

where C is a constant, Q is the gas flow rate (m³/s), α is the gas volume fraction in the rising plume and g is the acceleration due to gravity (m/s²); H and D are the same as in Equation (1.27). The constant C was evaluated (equal to 5.5×10^{-3}) by comparing the average spatial values of μ_e obtained from the $k-\varepsilon$ model [34]. The $k-\varepsilon$ model alone, when applied to gas-stirred ladle systems [35–36], overestimates the turbulence parameters. The value of C has been re-estimated from cold model experiments as 4.9×10^{-4} and the following equation has been proposed to estimate the average effective viscosity [35]:

$$\mu_e = 4.9 \times 10^{-4} \rho_l (gQ/D)^{1/3} \tag{1.29}$$

where ρ_l, g, Q and D are the same as described in Equation (1.28).

Simplifying assumptions

The assumptions made in describing the plume region are varied [36–38]. In an earlier work [37], the plume was assumed to be a solid core having only an upward velocity component. The measured velocity of the plume was taken as the boundary condition at the interface between the plume and the bulk. In another work [30] the plume region was considered to be a liquid of variable density depending upon the gas volume fraction. The slip between the liquid and gas bubbles was assumed to be zero and the gas volume fraction was calculated as the ratio of gas flow rate to overall gas–liquid volume flow rate in the plume region. A partial slip between the gas bubbles and the liquid was considered in the drift-flux model [36].

Empirical correlations have been developed (on the basis of water model experiments) to predict the void fraction distribution in the plume region [29]. When these empirical correlations were used to compute the velocity field and the map of the turbulent kinetic energy, the results were found to be in better agreement with the experiment than when the drift-flux [36] or no-slip [30] models were employed. The plume shape has also been predicted by considering the direct influence of bubble transport on the turbulence field, as well as by taking into account the additional generation of turbulence energy at the liquid–gas interface [39]. The theoretical predictions agreed well with the experimental results in air–water models [29]. The calculations, when extended to liquid steel, predicted the plume to be narrower than observed in air–water models. Hence, some tuning of the empirical correlations is required in extending the results of air–water models to the real processes. This aspect has also been demonstrated by experiments on an industrial scale with liquid steel stirred by gas in ladles [24].

1.4.2 *Circulation models*

Energy balance in a gas-stirred system can be simplified by assuming that, at a steady state, the sum of energy dissipation in liquid circulation ($\dot{\varepsilon}_c$) and bubble slip ($\dot{\varepsilon}_{bs}$) is equal to the rate of input energy, i.e. the stirring power of gas ($\dot{\varepsilon}$):

$$\dot{\varepsilon} = \dot{\varepsilon}_c + \dot{\varepsilon}_{bs} \tag{1.30}$$

The method of calculating $\dot{\varepsilon}$ for a given system is explained in Example 1.4.

In the circulation model of energy balance [40] the plume zone is assumed to be cylindrical in shape. The liquid and bubble velocities in the plume zone and the down-flow liquid velocity in the surrounding annular zone were considered to be uniform. The contribution of jet energy to mixing was neglected. In contrast to Equation (1.17), the mixing time, τ, was shown to depend not only on $\dot{\varepsilon}$ but also on system geometry:

$$\tau = 100[(D^2/H)^2/\dot{\varepsilon}]^{0.337} \tag{1.31}$$

where H and D are the height and diameter (in metres) of the bath, respectively. This expression was verified in several experiments [26]. Further improvements in the circulation model have been carried out to incorporate the following:

1. Shape of the plume zone
The shape of the plume zone is assumed to be conical (Figure 1.4) rather than cylindrical. Also, the plume angle depends on the operating variables.

2. Kinetic energy of the gas jet at the nozzle exit and the energy loss due to bubble break-up at the plume surface
The contribution of the kinetic energy of the jet to mixing depends upon operating variables such as gas flow rate, bath height and nozzle diameter and can amount to as much as 15% of the input energy [41]. The energy associated with the volume of liquid ejected due to bubble break-up on the bath surface in the plume region should also be considered.

3. Dimensionless mixing time
Since the mixing time is a function of vessel geometry, it is preferable to use a dimensionless mixing time or circulation number.

4. Delayed mixing due to the existence of the less turbulent and dead zones outside the plume region
The fluid flow in the bath, in the regions away from the plume and especially near the corners (dead zones) of the vessel, is not uniform. Lower fluid velocities or circulation in the less turbulent and dead zones thus affect the overall mixing process.

These modifications were incorporated in the lumped-parameter model [41] and the following correlations were obtained:

$$\phi = 0.028\dot{\varepsilon}^{0.209}H^{-0.825}d_n^{0.056} \tag{1.32}$$

$$U_p = 0.446\dot{\varepsilon}^{0.174} \tag{1.33}$$

and

$$\bar{U} = 2.81 \times 10^{-3}\dot{\varepsilon}^{0.625}h^{0.942}d_n^{0.119} \tag{1.34}$$

where ϕ is the dimensionless gas hold-up, $\dot{\varepsilon}$ is the rate of kinetic energy dissipation $(\text{kg m}^2/\text{s})$, H is the bath height (m), d_n is nozzle diameter (mm), U_p is the average liquid velocity inside the plume (m/s), $\bar{U}$ is average liquid recirculatory flow rate (m/s) and h is the plume height (m), defined as

$$h = H + h_d \tag{1.35}$$

where h_d is the projected plume section above the bath surface (m). The circulation time, t_c, is defined as

$$t_c = \frac{\text{volume of liquid in the mixing zone}}{\text{volumetric flow rate of the liquid in the mixing zone}}$$

$$= \frac{A_{op}H}{\bar{U}} = 83.1\dot{\varepsilon}^{-0.664}d_n^{-0.071} \tag{1.36}$$

where A_{op} is the cross-sectional area of the plume.

The circulation number, c_i, is defined as the number of circulations a liquid element makes in the vessel before attaining a certain desired degree of mixing. Mathematically,

$$c_i = \tau/t_c \tag{1.37}$$

c_i is correlated to the modified Froude number (Fr') and the geometrical parameters:

$$c_i = (Fr', H/D, d_n/D) \tag{1.38}$$

The regression equation to calculate c_i is given as

$$\ln c_i = 3.92 + 0.166 \ln Fr' - 6.87 \times 10^{-4} (\ln Fr')^2$$

$$- 0.636 \ln(H/D) - 0.138 \ln(H/D)^2$$

$$+ 0.395 \ln(d_n/D) - 0.037 \ln(d_n/D)^2 \tag{1.39}$$

The delay in mixing due to the existence of less turbulent and dead zones has been accounted for in partial volume models with two or more tanks [42–46], as well as the combined two-tank and recirculation model [47]. Besides these, the effective diffusion coefficient in the region outside the plume has been investigated [48]. Flow field and mixing with eccentric gas stirring have been investigated by employing a multitank model connected in series [49]. It is found in practice that the dead zones occurring at the bottom of the vessel in axisymmetric gas injection can be avoided by using an eccentrically placed nozzle (or tuyère) at the bottom. The mixing time thus decreases.

1.4.3   *Empirical models*

Dimensional analysis of the fluid flow based on the Navier–Stokes equation has been carried out [50] to determine the relationships between mixing power density (defined as $\dot{\varepsilon} = dW/dt$ where W is work done), characteristic fluid velocity, u (m/s), characteristic length, L (m), mixing length, l (m) and viscosity, μ (kg/m s) as follows:

For the fluid motion dominated by viscous force:

$$u \propto (L^2 \dot{\varepsilon}/\mu)^{1/2} \tag{1.40}$$

For the fluid motion dominated by the force of inertia:

$$u \propto (L \dot{\varepsilon}/\mu)^{1/3} \tag{1.41}$$

For the fluid motion dominated by turbulent force:

$$u \propto (L^3 \dot{\varepsilon}/\mu)^{1/3} \tag{1.42}$$

The exponents 1/2 and 1/3 in Equations (1.40)–(1.42) have been verified in a large number of experiments reported in the literature [50].

The dimensionless product group P defined for the homogeneous buoyant jets has been adapted [51] to describe the heterogeneous buoyant jet:

$$P = g^{-1/5}(\varepsilon_1 Q)^{2/5} x^{-1} \tag{1.43}$$

where Q is gas flow rate, x is the vertical coordinate, ε_1 is the dimensionless density (thermal plume versus bulk) and g is acceleration due to gravity; in the case of pure liquid $\varepsilon_1 = 1$. The equations used to describe the homogeneous buoyant jet were modified to describe the heterogeneous buoyant plume. That is, for the kinematic flow rate of buoyancy ($\dot{W}$), radius (r), axial velocity (u_o), volumetric flow capacity ($\dot{V}$) and the dimensionless density difference (ε_o), suitable exponents n_1, n_2, n_3 and n_4 and constants a_1, a_2, a_3 and a_4 were used as follows:

Homogeneous buoyant jet	Heterogeneous buoyant plume	
$r = 0.134x$	$r = a_1 p^{n_1} x$	(1.44)
$u_o = 3.79 \dot{W}_1^{1/3} x^{-1/3}$	$u_o = a_2 p^{n_2} \dot{W}_1^{1/3} x^{-1/3}$	(1.45)
$\dot{V} = 0.216 \dot{W}_1^{1/3} x^{5/3}$	$\dot{V} = a_3 p^{n_3} \dot{W}_1^{1/3} x^{5/3}$	(1.46)
$\varepsilon_o = 11 p^{5/3}$	$\varepsilon_o = a_4 p^{n_4}$	(1.47)

The values of the constants a_1, a_2, a_3 and a_4 and the exponents n_1, n_2, n_3 and n_4 were then determined by empirically fitting (regression analysis) the results obtained from water model experiments:

$$r = 0.38Q_1^{0.15}x^{0.62} \tag{1.48}$$

$$u_o = 3.37Q_1^{0.25}x^{-0.12} \tag{1.49}$$

$$\dot{V} = 1.52Q_1^{0.55}x^{1.13} \tag{1.50}$$

In the case of liquid steel at temperature T_l, the value of Q is modified as follows:

$$Q_1 = Q_N \frac{T_l}{T_N}\left(\frac{h_N}{h_a + H}\right) \tag{1.51}$$

where the subscript N refers to standard conditions of temperature and pressure, and H is liquid depth, h_a is pressure height at $x = H$ and h_N is pressure height under standard conditions for the gas phase.

Based on laboratory experiments, many empirical correlations have been proposed for the speed of liquid recirculation and the rising velocity of water in the plume. For example, the mean rising velocity of water in the plume created by bubbling jets is given by [52]

$$\bar{U}_p = 0.915Q^{0.225} \tag{1.52}$$

Similarly, the following empirical relations have been proposed to predict the ratio of the mean speed of liquid recirculation ($\bar{U}$) to the average plume velocity (U_p) on the basis of water model experiments [38]:

$$\frac{\bar{U}}{U_p} = \frac{0.18}{R_v^{1/3}} \tag{1.53}$$

where R_v is vessel radius (m). The centreline velocity and gas volume fraction were evaluated with the help of Equation (1.53) and then used to estimate buoyancy. The average plume velocity (U_p) can be estimated [35] from

$$U_p = 4.78\frac{(Q^{1/3}L^{1/4})}{R_v^{1/4}} \tag{1.54}$$

where Q is the gas flow rate (m^3/s) and L is liquid depth (m). On substituting U_p from Equation (1.54) in Equation (1.53), the expression for the mean speed of liquid recirculation can be obtained as

$$\bar{U} = \frac{(0.86Q^{1/3}L^{1/4})}{R_v^{7/12}} \tag{1.55}$$

Several other studies [53, 54] have been reported to describe the plume in air–water models. Plume characteristics and liquid circulation caused by gas injection through a porous plug have also been investigated [55].

Example 1.3 **Average plume velocity and mean speed of liquid recirculation in a gas-stirred vessel**

Given the following for an oxygen steelmaking converter:

- inert gas flow rate through the bottom, $Q = 0.0416 \text{ m}^3/\text{s}$
- liquid depth, $L = 1.85 \text{ m}$
- vessel radius, $R_v = 3.2 \text{ m}$

Substitution of Q, L and R_v in Equation (1.54) gives the average plume velocity as $U_p = 1.44 \text{ m/s}$.

On substituting U_p and R_v in Equation (1.53), the mean speed of liquid recirculation can be calculated as $\bar{U} = 0.18 \text{ m/s}$.

It is obvious that if the vessel has a large radius, or if it is oblong (e.g. in the case of a torpedo used to transport hot metal in a steel plant), then the flow rate of the liquid metal from the far end of the vessel to the plume, where gas is rising and mixing or chemical reactions are taking place, will be slow. In order to improve mixing and mass transfer, therefore, several tuyères are dispersed uniformly through the bath in oxygen steelmaking converters instead of using a single tuyère (or nozzle) at the centre of the vessel. For the same reason, it is preferable to use a ladle rather than a torpedo, if desulphurization is to be carried out in the vessel by gas-borne injection of a powder into liquid metal.

Example 1.4 **Energy dissipation and mixing time in a gas-stirred bath when tuyères are placed symmetrically at the bottom**

If the bottom of the vessel is fitted with one tuyère, then the mixing time can be calculated from Equation (1.17). If more than one tuyère is used, then Equation (1.17) is modified as

$$\tau = 800\dot{\varepsilon}^{-0.40}N^{1/3} \tag{A}$$

The theoretical value of $\dot{\varepsilon}$ can be calculated from

$$\dot{\varepsilon} = \left(28.5 \frac{Q_B T}{W}\right)\lg\left(1 + \frac{H}{148}\right) \tag{B}$$

Suppose for a 300 ton steelmaking vessel that the different parameters of Equations (A) and (B) are

Q_B = flow rate of bottom gas, $15 \text{ m}^3 \text{ (stp)/min}$

N = number of stirring elements, 6

W = bath weight, 300 tons

H = bath depth, 185 cm

$T = 1873 \text{ K}$

D = bath diameter, 630 cm

On substituting the values of Q_B, W, H and T into Equation (B):

$$\dot{\varepsilon} = \left(28.5 \, \frac{15 \times 1873}{300}\right) \lg\left(1 + \frac{185}{148}\right) \tag{B}$$

$$= 940 \text{ W/ton}$$

On substituting the values of $\dot{\varepsilon}$ and N into Equation (A):

$$\tau = 800 \times 940^{-0.40} 6^{1/3} = 94 \text{ s}$$

If the number of tuyères is reduced to four but the gas flow rate is kept the same, then $\tau = 82.13$ s. Similarly, if only one tuyère is used, $\tau = 51.7$ s.

The mixing time can also be calculated from Equation (1.31) which is valid only for the case of a single tuyère. On substituting the values in Equation (1.31):

$$\tau = 100\left[\left(\frac{6.3^2}{1.85}\right)^2 \left(\frac{1}{940}\right)\right]^{0.337}$$

$$= 73.6 \text{ s}$$

Equation (1.31) predicts that the mixing time decreases if bath height is increased.

Mixing time is a good indicator of the temperature and concentration homogenization of the bulk of liquid metal but we should not confuse it with the mass transfer phenomena occurring at the slag–metal interface. In fact, for the same flow rate of gas in a combined blown converter, opposite trends can be observed as far as the effect of number of tuyères on mixing and interfacial mass transfer is concerned.

1.5 Impinging gas jets

The application of impinging, supersonic, pure oxygen jets to oxidize the dissolved impurities in liquid iron alloys led to the invention, in the late 1950s in Austria, of the LD process (Linz–Donawitz process). In this process a major portion of the kinetic energy of the jet impinging upon liquid metal is used to eject metal droplets which, in turn, provide a large surface area for the oxidation–reduction reactions at the interface of gas, slag and metal phases. The thermodynamic and kinetic aspects of the reactions in the LD process are discussed in Chapters 5–8. This section is devoted to the fundamentals of gas flow through nozzles. Other physical effects of the jet–metal interaction are discussed in Chapters 5 and 6.

1.5.1 *Impinging supersonic gas jets*

A properly designed nozzle, fitted at the end of a pipe, can accelerate oxygen gas to supersonic velocity by converting the pressure energy of the gas into kinetic energy.

Owing to the compressible nature of the gas phase and the short length of the nozzle, the frictional losses are small and the entropy of the gas remains constant (i.e. isentropic flow). A steady gas flow through the nozzle can thus be studied by assuming that the flow variables – gas pressure (p), density (ρ) and velocity (u) in the direction of the flow – are constant at each cross-section (A) of the nozzle and vary only along the length (i.e. one-dimensional flow). Since the mass of the gas (W) at each point along the flow is constant

$$W = \rho u A = \text{constant}$$

or

$$\nabla \rho u A = 0$$

or

$$\frac{d\rho}{\rho} + \frac{du}{u} + \frac{dA}{A} = 0 \quad \text{(equation of continuity)} \tag{1.56}$$

For an ideal gas

$$p = \rho R T \tag{1.57}$$

where R is the gas constant and T is the gas temperature. In the case of a reversible, adiabatic process, it can be shown that

$$p/\rho^\gamma = \text{constant} \tag{1.58}$$

where γ is the ratio of the specific heats of the gas at constant pressure (C_p) and at constant volume (C_v):

$$C_p/C_v = \gamma \tag{1.59}$$

The value of γ for oxygen is 1.4. Also, for an ideal gas,

$$C_p - C_v = R \tag{1.60}$$

Logarithmic differentiation of Equation (1.58) and then using Equation (1.57) gives

$$\frac{dp}{d\rho} = \frac{\gamma p}{\rho} = \gamma R T \tag{1.61}$$

In an adiabatic process, the speed at which the compression–expansion wave passes through a medium (i.e. the speed of sound, s) is related to the pressure and density changes according to

$$s = \left[\left(\frac{\partial p}{\partial \rho} \right) \right]^{1/2}_{\text{const. entropy}} \tag{1.62}$$

Hence from Equation (1.61)

$$s = (\gamma R T)^{1/2} \tag{1.63}$$

The Bernoulli equation for frictionless flow in one dimension is

$$u\nabla u + \frac{1}{\rho}\nabla p = 0 \quad \text{or} \quad u\,\mathrm{d}u + \frac{1}{\rho}\,\mathrm{d}p = 0 \tag{1.64}$$

Substituting for $\mathrm{d}p$ from Equation (1.62) in Equation (1.64)

$$u\,\mathrm{d}u + s^2\frac{\mathrm{d}\rho}{\rho} = 0 \tag{1.65}$$

Eliminating $\mathrm{d}\rho/\rho$ between Equations (1.56) and (1.65)

$$\left(1 - \frac{u^2}{s^2}\right)\frac{\mathrm{d}u}{u} + \frac{\mathrm{d}A}{A} = 0 \tag{1.66}$$

In a high-velocity flow, the ratio of gas velocity to the speed of sound at any given point is defined as the dimensionless Mach number, Ma:

$$Ma = u/s \tag{1.67}$$

Substituting in Equation (1.66):

$$\frac{\mathrm{d}A}{A} = (Ma^2 - 1)\frac{\mathrm{d}u}{u} \tag{1.68}$$

This equation shows that for $0 < Ma < 1$, i.e. subsonic flow, a decrease in area (or negative value of $\mathrm{d}A/A$) will lead to an increase in velocity or a positive value of $\mathrm{d}u/u$. For supersonic flow, or $Ma > 1$, a positive value of $\mathrm{d}A/A$, or an increase in area, will also lead to an increase in velocity. At $Ma = 1$, i.e. sonic flow, $\mathrm{d}A = 0$. This can be physically interpreted by saying that for a continuous acceleration of gas from subsonic to supersonic speeds, the flow must pass through a minimum area. Convergent–divergent nozzles are called de Laval nozzles. The only physical restriction is that the gas pressure at the exit of the nozzle (p_e) must be slightly greater than the ambient pressure. Otherwise, a pressure-equalizing mechanism develops and shock waves are produced. As a result of this effect, part of the jet energy is lost even before the jet strikes the liquid metal surface. Usually, a multiple nozzle lance tip (more than one nozzle at the tip of the lance) is used in steelmaking converters. In this chapter only the single-hole de Laval nozzle is considered.

The total specific enthalpy of gas, H_t, at any point in the flow is equal to the sum of the specific enthalpy H and the kinetic energy of unit mass of gas at that point:

$$H_t = H + \tfrac{1}{2}u^2 \tag{1.69}$$

Referring to Figure 1.5, consider any two points (1 and 2) in the flow field with flow variables (p_1, ρ_1, T_1) and (p_2, ρ_2, T_2), respectively. The corresponding values in the pipe (upstream), at the throat of the nozzle and at the nozzle exit are: (p_o, ρ_o, T_o), (p_t, ρ_t, T_t) and (p_e, ρ_e, T_e), respectively.

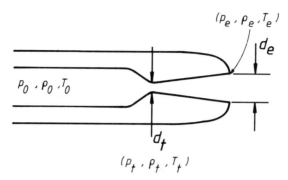

Figure 1.5   Pressure density and temperature variables at different points along the length of a de Laval nozzle. (*Source*: Geiger, G. H. and Poirier, D. R. (1973) *Transport Phenomena in Metallurgy*, Addison-Wesley: Reading, MA, p. 162.)

Since energy is conserved, the generalized equation of energy balance at any two points with enthalpy and velocity values (H_1, u_1) and (H_2, u_2) gives

$$H_1 + \tfrac{1}{2}u_1^2 = H_2 + \tfrac{1}{2}u_2^2 = \text{constant} \tag{1.70}$$

For a constant value of C_p, $H = C_p T$. Substituting in Equation (1.70):

$$C_{p_1} T_1 + \tfrac{1}{2}u_1^2 = C_{p_2} T_2 + \tfrac{1}{2}u_2^2 = \text{constant} \tag{1.71}$$

From Equations (1.59) and (1.60):

$$C_p = \frac{\gamma R}{\gamma - 1} \tag{1.72}$$

From Equations (1.72), (1.63) and (1.67), Equation (1.71) becomes

$$T_1\left(1 + \frac{\gamma - 1}{2} Ma_1^2\right) = T_2\left(1 + \frac{\gamma - 1}{2} Ma_2^2\right) \tag{1.73}$$

where $Ma_1 = u_1/s$ and $Ma_2 = u_2/s$. Since $p = \rho R T$ and R is constant throughout the flow region:

$$\frac{p_1 \rho_2}{p_2 \rho_1}\left(1 + \frac{\gamma - 1}{2} Ma_1^2\right) = \left(1 + \frac{\gamma - 1}{2} Ma_2^2\right) \tag{1.74}$$

Substituting for pressure from Equation (1.58) into the above equation:

$$\rho_1^{\gamma - 1}\left(1 + \frac{\gamma - 1}{2} Ma_1^2\right) = \rho_2^{\gamma - 1}\left(1 + \frac{\gamma - 1}{2} Ma_2^2\right) \tag{1.75}$$

By following a similar procedure, ρ can be eliminated and

$$p_1^{(\gamma - 1)/\gamma}\left(1 + \frac{\gamma - 1}{2} Ma_1^2\right) = p_2^{(\gamma - 1)/\gamma}\left(1 + \frac{\gamma - 1}{2} Ma_2^2\right) \tag{1.76}$$

In the above treatment, the points 1 and 2 are completely arbitrary in the flow field. Particular relations between upstream conditions (p_o, ρ_o, T_o), at the throat (p_t, ρ_t, T_t), and at the nozzle exit (p_e, ρ_e, T_e) can now be obtained.

The Mach number of the gas in the pipe (upstream) can be considered to be approximately zero because the cross-section of the pipe is much larger than that of the nozzle. This is also referred to as the stagnation condition in the pipe. The relation (1.76) can be written as

$$p_o^{(\gamma-1)/\gamma}\left(1 + \frac{\gamma-1}{2} Ma_o^2\right) = p_2^{(\gamma-1)/\gamma}\left(1 + \frac{\gamma-1}{2} Ma_2^2\right)$$

and since $Ma_o = 0$:

$$p_o^{(\gamma-1)/\gamma} = p_2^{(\gamma-1)/\gamma}\left(1 + \frac{\gamma-1}{2} Ma_2^2\right) \tag{1.77}$$

Similarly

$$T_o = T_2\left(1 + \frac{\gamma-1}{2} Ma_2^2\right) \tag{1.78}$$

$$\rho_o^{\gamma-1} = \rho_2^{\gamma-1}\left(1 + \frac{\gamma-1}{2} Ma_2^2\right) \tag{1.79}$$

From Equations (1.77)–(1.79), we can easily obtain

$$\frac{p_2}{p_o} = \left(\frac{\rho_2}{\rho_o}\right)^\gamma = \left(\frac{T_2}{T_o}\right)^{\gamma/(\gamma-1)} \tag{1.80}$$

If point 2 refers to the throat where the Mach number is unity and $\gamma = 1.4$ for oxygen, then Equations (1.77)–(1.79) can be written as

$$p_t/p_o = 0.5283 \tag{1.81}$$

$$T_t/T_o = 0.823 \tag{1.82}$$

and

$$\rho_t/\rho_o = 0.6339 \tag{1.83}$$

The p_2/p_o values calculated from Equation (1.76) are plotted against Mach number in Figure 1.6. At the throat the value of p_t/p_o is 0.5283. In Equations (1.76)–(1.79), the point 2 can be replaced by flow parameters at the nozzle exit (p_e, ρ_e, T_e). The nozzles in impinging jet steelmaking are designed to operate in a Mach number range of 2–2.4 at the nozzle exit (see Figure 1.6).

Since $Ma = u/s$, $s = (\gamma RT)^{1/2}$ and $R_u = R(MO_2)$, where R_u is the universal gas constant and MO_2 is the molecular weight of oxygen gas, the velocity of the gas can

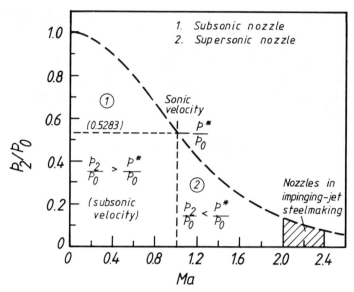

Figure 1.6   Classification of gas injection nozzles according to Mach number. (*Source*: Koria, S. C. (1988) *Steel Res.*, **59**, 257–62.)

be obtained from Equation (1.76) as

$$u_2 = \left\{ \left(\frac{2\gamma}{\gamma - 1} \right) \left(\frac{p_0}{\rho_0} \right) \left[1 - \left(\frac{p_2}{p_0} \right)^{(\gamma - 1)/\gamma} \right] \right\}^{1/2}$$

$$= \left\{ \left(\frac{2\gamma}{\gamma - 1} \right) \left(\frac{R_u T_0}{MO_2} \right) \left[1 - \left(\frac{p_2}{p_0} \right)^{(\gamma - 1)/\gamma} \right] \right\}^{1/2} \tag{1.84}$$

This expression can be used to calculate the exit velocity u_e by substituting $p_2 = p_e$. If the Mach number at the exit is Ma_e, it can be shown from Equation (1.76) or from Equation (1.84) that

$$Ma_e^2 = \frac{2}{(\gamma - 1)} \left[\left(\frac{p_0}{p_e} \right)^{(\gamma - 1)/\gamma} - 1 \right] \tag{1.85}$$

In a convergent–divergent nozzle (Figure 1.5), if the area of the throat is A_t and the area at the exit is A_e, then the constant mass flow rate ($\dot{m}$) of gas is given by

$$\dot{m} = A_e \rho_e u_e = A_t \rho_t u_t \tag{1.86}$$

For sonic velocity at the throat ($Ma = 1$), it can be shown that

$$\frac{A_t}{A_e} = \left(\frac{\gamma + 1}{2} \right)^{1/(\gamma - 1)} \left\{ \left(\frac{\gamma + 1}{\gamma - 1} \right) \left[\left(\frac{p_e}{p_0} \right)^{2/\gamma} - \left(\frac{p_e}{p_0} \right)^{(\gamma + 1)/\gamma} \right] \right\}^{1/2} \tag{1.87}$$

The value of mass flow rate parameter [56], defined as $((\dot{m}/Ap_0)\sqrt{T})$, is a maximum (see Figure 1.7) when (p/p_0) corresponds to the critical pressure ratio given

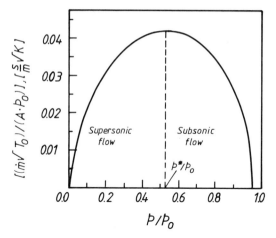

Figure 1.7   Mass flow characteristics of oxygen gas through a duct of varying cross-section. (*Source*: Koria, S. C. (1988) *Steel Res.*, **59**, 104–9.)

by Equation (1.81). At this pressure ratio the nozzle is said to be choked, i.e. any further increase in the upstream pressure does not cause an increase in the mass flow rate of the gas. The maximum mass flow rate ($\overset{*}{m}_t$), on substituting the requisite values in Equation (1.86), is given by

$$\overset{*}{m}_t = A_t\left[\rho_o p_o \gamma \left(\frac{2}{\gamma + 1}\right)^{(\gamma + 1)/(\gamma - 1)}\right]^{1/2} \tag{1.88}$$

Example 1.5 **Dimensions of a supersonic nozzle for an oxygen steelmaking converter**

The procedure for calculating the nozzle dimensions can be demonstrated as follows. Suppose a nozzle is to be designed to achieve Mach 2 at the nozzle exit when the oxygen flow rate is 400 m³ (stp)/min and the exit pressure, p_e, is 1 bar. Substituting $M_e = 2$, $p_e = 1$ and also $\gamma = 1.4$ (for oxygen) in Equation (1.85), the driving or upstream pressure, p_o, can be calculated as $p_o = 7.824$ atm.

The ratio of the throat area to the exit area can be calculated from Equation (1.87):

$$A_t/A_e = 0.591$$

The throat area can be calculated from Equation (1.88) by substituting the values as follows:

$$\overset{*}{m}_t = \frac{400 \times 32}{22.4 \times 60} = 9.524 \text{ kg/s}$$

$$\rho_o = \frac{p_o M O_2}{R_u T_o} = \frac{7.824 \times 1.013 \times 10^5 \times 32}{8314 \times 298} = 10.224 \text{ kg/m}^3$$

Hence A_t from Equation (1.88) can be calculated as $4.886 \times 10^{-3}\,\text{m}^2$ or the throat diameter, d_t, is 0.0789 m. Since the ratio of A_t/A_e, as calculated above, is 0.591 the exit area is $8.267 \times 10^{-3}\,\text{m}^2$ or the exit diameter is 0.1026 m.

In the above calculations, the exit pressure has been assumed to be 1 atm. The exit pressure should always be slightly greater than the ambient pressure, otherwise there will be a tendency for the dust-laden gases to be sucked in from the outer periphery of the nozzle walls into the nozzle (causing nozzle wear). An additional requirement for the multiple nozzle lance tips is that the jet produced from each nozzle should not interfere with the jets from other nozzles, i.e. non-coalescing jets. This is achieved by placing each nozzle at an angle from the vertical axis.

1.5.2   *Decay of supersonic gas jets*

The supersonic jet issuing from the nozzle has a supersonic core in which the gas velocity is higher than Mach 1. The diameter of the core gradually narrows down (Figure 1.8). For the nozzles used in oxygen steelmaking, the length of the supersonic core usually varies from three to eight times the nozzle exit diameter. After the decay

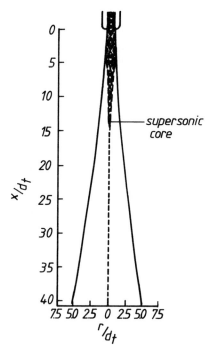

Figure 1.8   Decay of supersonic jet. (*Source*: Anderson, A. R. and Johns, F. R. (1955) *Jet Propulsion*, **25**, 13–15, 25.)

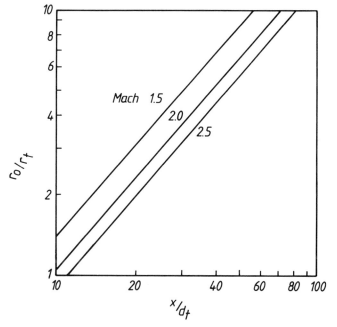

Figure 1.9 Jet-spreading characteristics as a function of Mach number. (*Source:* Anderson, A. R. and Johns, F. R. (1955) *Jet Propulsion*, **25**, 13–15, 25.)

of the supersonic core, the jet begins to spread by entrainment of the air (or gases) from the ambient atmosphere. The overall spreading of the jet is expressed either as the ratio of velocity (u_r) at radius r to the centreline velocity (u_o), or by the ratio r_o/r_t, where r_o is the radius at which the velocity is one-half of the velocity at the centreline and r_t is the radius of the throat of the nozzle. The jet-spreading characteristics as a function of Mach number and the distance from the nozzle exit are shown in Figure 1.9. Beyond the length of the supersonic core, the jet expands at an included angle of approximately 18°. The effective jet radius at any location can be taken as $2r_o$. Thus, for a given value of x/d_t, Mach number and r_t, the effective jet radius ($2r_o$) can be obtained from Figure 1.9.

The effects of entrainment of the gases from the surrounding atmosphere into the jet on jet decay characteristics, on the recirculation of gases and the effect of jet impact on metal droplet generation, etc., are discussed in detail in Chapter 5 (devoted to slag formation).

1.6 Physical modelling

A large number of experiments are required to understand the nature of interaction between different phases in an industrial-scale steelmaking process. It is much easier, safer and less expensive, as discussed in section 1.4.1, to conduct experiments first on a laboratory or pilot scale and then on an industrial scale. However, in order to apply the findings of the laboratory or pilot-scale experiments to industrial conditions, it is necessary to establish appropriate similarity criteria between the experiment on the small scale and the practice in the industry.

The types of similarities which can be maintained between two systems and their basic operating units are as follows:

Type of similarity	Basic operating units
1. Geometric	Length
2. Mechanical	
Static	Length, force
Kinematic	Length, force, time
Dynamic	Length, force, time
3. Magnetic	Magnetic energy
4. Electric	Electric energy
5. Thermal	Length, force, time, temperature
6. Chemical	Mass transfer

Geometric similarity refers to the correspondence in physical shape and size of systems, i.e. same length to height ratio etc. Mechanical similarity comprises static, kinematic and dynamic similarities. The static similarity criterion is concerned with solid bodies and structures and is not relevant to steelmaking systems. Similarly, the electric and magnetic similarity criteria are also not relevant to steelmaking (except in very special cases like electromagnetic stirring etc.). Kinematic similarity requires that the corresponding particles in the model and in the prototype trace out geometrically similar paths in corresponding intervals of time. Dynamic similarity is essentially concerned with forces acting on a system. Let the forces, for example, be as follows (adapted from Guthrie [61]):

Pressure force, F_p = pressure × cross-sectional area, pL^2

Inertial force, F_i = mass × acceleration, $\rho u^2 L^2$

Gravity force, F_g = weight of fluid, $\rho g L^3$

Viscous force, F_μ = tangential shear stress × area. $\mu u L$

For the above forces acting on a system, the dynamic similarity between a model (m) and a prototype (p) requires that at a given point

$$\frac{F_{p,m}}{F_{p,p}} = \frac{F_{i,m}}{F_{i,p}} = \frac{F_{g,m}}{F_{g,p}} = \frac{F_{\mu,m}}{F_{\mu,p}} = C_F$$

where C_F is a constant. It follows that the ratio of different forces between model and prototype must be also equal, for example

$$\frac{F_{i,m}}{F_{\mu,m}} = \frac{F_{i,p}}{F_{\mu,p}}$$

Or,

$$\left(\frac{\rho u^2 L^2}{\mu u L}\right)_m = \left(\frac{\rho u^2 L^2}{\mu u L}\right)_p$$

which means that the Reynolds number, Re for the model and the prototype must be equal:

$$Re_m = Re_p$$

Similarly, from the ratio of inertia and gravity forces the Froude number must be same, i.e.

$$Fr_m = Fr_p$$

and the pressure drop along a flow system must also be the same, i.e. the Euler number for the model and the prototype are equal:

$$Eu_m = Eu_p$$

Thermal similarity requires that the rate of heat transfer by various mechanisms (i.e. by conduction, radiation and convection) must be of a fixed ratio for the model and the prototype.

In chemically similar systems the concentration difference at corresponding geometrical points is similar (i.e. in a constant ratio). In steelmaking, since most reactions are controlled by mass transfer, the chemical similarity criterion can be easily satisfied in a prototype provided the geometric, dynamic and kinematic similarities are established.

In addition to the above similarity criteria, geometrical scaling factors are used in order to correlate the different scales of experimentation. For example, mixing time in a ladle is defined as the time corresponding to an arbitrary level of approach (say 95%) to the final steady-state concentration. The geometrical scale λ (i.e. the ratio of the height of a full-size ladle to that of the model employed in the laboratory) is used to take into account the effect of size of experimentation in the laboratory in comparison with industry:

$$(\tau)_{model} = (\tau)_{full\ scale}\ \lambda^n \tag{1.89}$$

where τ indicates mixing time. In the case of axisymmetrically gas-agitated systems [57] the value of n is $3/4$; if the metal is covered by a layer of slag, the value of n increases. In order to decide the minimum number of parameters to be studied in a system, while fulfilling the criteria of similarity, it is customary to employ the Buckingham (pi) theorem [58] (see the worked example in Chapter 3).

The concepts of physical modelling and the appropriateness of the similarity criteria invoked in an experimental set-up will become clearer from the two examples discussed below.

Example 1.6 Cold model study of an argon-stirred ladle

In the physical modelling of an argon-stirred ladle we should establish, in principle, the conditions of geometric, dynamic, thermal and chemical similarities between the experimental set-up in the laboratory and the actual process on the shop floor.

Let us analyze the cold model work to study the effects of argon stirring in a 200 ton steel ladle on a laboratory scale. Only the fundamentals are explained here; for actual results reference must be made to Kim and Fruehan's paper [59].

A 1/7.2 scale model (geometric similarity) of the vessel was made. The bottom diameter of the experimental (cold model) vessel was 45.6 cm. The height was 62 cm with a 10% (2.5°) taper. The liquid was filled to a height of 44.5 cm. Four different types of tuyère positions were used to study the effect of tuyère location: one at the centre, one each at one-half the radius, three in a line including one at the centre and one each at one-half the radius, and one at the bottom of the vessel side wall. The tuyère diameter was either 0.2 cm or 0.48 cm. A stainless steel tube with side holes or ports (0.48 cm diameter) was used for submerged gas injection at different depths. The modified Froude number was considered as the similarity criterion to correlate the gas flow rate between the laboratory model and the full-scale system. The linear gas velocity (u_t) at the exit of the tuyère was calculated from

$$\rho_g u_t = \frac{Q}{22.4} \frac{M_g}{60} \frac{4}{\pi d_t^2}$$

where Q is gas flow rate (l/min), ρ_g is the density of gas (g/cm^3), M_g is the molecular weight of gas (g) and d_t is the diameter of the tuyère (cm). The modified Froude number (Fr') is given by

$$Fr' = \frac{\rho_g u_t^2}{(\rho_1 - \rho_g)gL} = C\frac{Q^2}{Ld_t^2}$$

where ρ_1 is the density of the liquid, g is acceleration due to gravity, L is bath depth and

$$C = \frac{9.159 \times 10^{-10} M_g^2}{(\rho_1 - \rho_g)}$$

Since $(Fr')_m = (Fr')_{FS}$, where the subscript 'm' refers to model and 'FS' to full scale,

$$\frac{C_m Q_m^2}{L_m (d_t^2)_m} = \frac{C_{FS} Q_{FS}^2}{L_{FS}(d_t^2)_{FS}}$$

or

$$Q_{FS}^2 = \frac{C_m}{C_{FS}} Q_m^2 \left(\frac{L_{FS}}{L_m}\right) \frac{(d_t^2)_{FS}}{(d_t^2)_m}$$

If λ is the geometrical scaling factor:

$$\lambda = L_{FS}/L_m$$

The following relation was used in the cold model work [59]:

$$Q_{FS} = 1.038\lambda^{3/2}Q_m$$

for geometrical scaling down from a steel–argon ladle (full scale) to a water/air model on a laboratory scale. A gas flow of 28.32 l/min in the 200 ton ladle thus corresponded to 0.21 l/min in the cold model experiments.

To establish chemical similarity with respect to reactions between the metal and slag, i.e. the transfer of sulphur from metal to slag, a 50–50 mixture (by volume) of paraffin oil and cotton seed oil was taken as a slag phase on top of water. Thymol ($C_{10}H_{14}O$) was used as the tracer. The equilibrium partition ratio of thymol in an oil/water system is approximately 300. The weight ratio of oil and water was similar to that of slag and steel (33 kg/ton). Some of the effects which could not be simulated were as follows:

- Being a cold model study the thermal effects could not be simulated.
- Interfacial tension in the cold model was different from that in a slag–metal system.
- Molecular diffusivities in the cold model were not comparable with those in metal and slag.
- Sulphur transfer is a coupled inorganic reaction of the type

  $$[S] + (CaO) \rightarrow (CaS) + [O]$$

  where the transfer of thymol from water to oil is an organic reaction.
- In the cold model experiment an attempt was made to simulate the metal side mass transfer only. In reality sulphur transfer can be mixed transport controlled.

Example 1.7 **Cold model study of refractory wear in oxygen steelmaking converters**

The wearing out of refractory bricks in oxygen steelmaking converters is a complex phenomenon. It is attributed to the combined effects of technological factors, refractory quality, design of the vessel, the brick-laying pattern used in the vessel and the blowing regime. The technological factors include silicon content of hot metal, lime quality, interheat delays, number of reblown heats, end point temperature of liquid steel, etc. The refractory quality depends upon bulk density, impurity levels of oxides of type R_2O_3, cold crushing strength, porosity, carbon content, loss on ignition, etc. The blowing regime essentially involves the adjustment of lance height depending upon nozzle design and bath depth.

A cold model investigation was carried out to study the effect of blowing regime in a top-blown 220 ton steelmaking converter [60]; only the fundamentals are explained here and for actual results reference must be made to the paper. The different similarity criteria were established as follows. A 1/20 scale model (geometric similarity) of the 220 ton refractory-lined converter was made. Wax was used as the lining material to represent the monolithic lining (refractory covered with a thin layer of slag) and kerosene represented the corrosive fluid (liquid slag). A straight-bore nozzle of 0.30 cm diameter was employed to blow air. In the actual converters a convergent–divergent nozzle (de Laval nozzle) is used. The configuration of the nozzle does not affect jet penetration as long as the lance height, nozzle diameter and blowing rate are kept the same. The lance height from the surface of the kerosene bath was determined from the dimensionless lance distance (h_{dl}) for the actual converter:

$$h_{dl} = h/nd_t$$

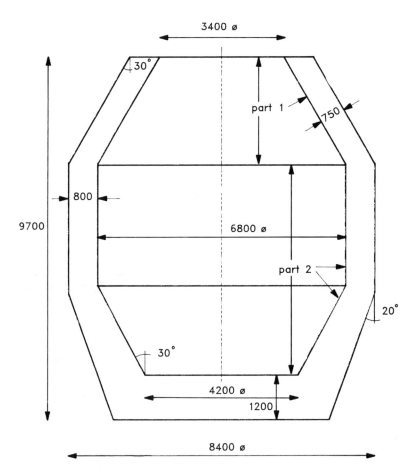

Figure 1.10 Dimensions of a 220 ton LD converter; all dimensions in millimetres.

where h is the actual lance height, n is the number of nozzles in the lance tip and d_t is the throat diameter for the actual converter (220 tons). The height of the liquid kerosene in the model was determined from the dimensionless ratio of bath depth and the maximum inner diameter of the bath for the actual converter. For example, according to the dimensions for a new converter lined with refractory (Figure 1.10), the bath depth and the maximum inner diameter may be taken as 2 m and 6.8 m, respectively. The dimensionless ratio is thus 0.29. As the refractory wear progresses, this ratio will change and therefore three different levels of bath depth were employed. The flow rate of air (or blowing pressure) was determined by equating the modified Froude number (Fr') for the model and the full scale. The major limitations of this study, as generally true for most of the cold experiments reported in the literature, were as follows:

- There was no temperature rise and hence no thermal effects (no thermal similarity).
- There were no such chemical reactions as those between the oxygen jet and the liquid metal in the jet impact zone and also the chemical attack of the slags of varying composition was not simulated (no chemical similarity).

References

1. Krieger, W. and Trenkler, H. (1971) 'The significance of melt structures of iron–carbon and iron–nickel alloys', *Arch. Eisenhüttenwes.*, **42**, 175–84.
2. Trenkler, H. and Krieger, W. (1978) In *Metallurgie des Eisens*, Vol. 7(a), *Theorie der Stahlerzeugung*, [Gmelin-Durrer] Springer: Berlin, pp. 24a–51a.
3. Keene, B. J. (1988) 'Review of data for the surface tension of iron and its binary alloys', *Int. Mater. Rev.*, **33** (complete issue devoted to the subject).
4. Kawai, Y. (1971) 'Viscosity and surface tension of liquid iron alloys', *Trans. Iron Steel Inst. Jpn*, **11**, Suppl. 1, 387–90.
5. Ogino, Y., Borgmann, F. O. and Frohberg, M. G. (1974) 'On the anomalous change of viscosity of liquid iron with temperature', *Trans. Iron Steel Inst. Jpn*, **14**, 82–7.
6. Mills, K. C. and Keene, B. J. (1987) 'Physical properties of BOS slags', *Int. Mater. Rev.*, **32** (complete issue devoted to the subject), 1–120.
7. *Slag Atlas* (1981) Verlag Stahleisen: Düsseldorf.
8. *Phase Diagrams for Ceramists* (ed. M. K. Reser), American Ceramic Society: Columbus, OH, 1964, 1969, 1975.
9. Turkdogan, E. T. (1983) *Physicochemical Properties of Molten Slags and Glasses*, The Metals Society: London.
10. Koch, K. and Janke, D. (1984) *Schlacken in der Metallurgie*, Verlag Stahleisen: Düsseldorf.
11. Harlow, F. H. and Nakayama, P. I. (1967) 'Turbulent transport equations', *Phys. Fluids*, **10**, 2323.
12. Launder, B. E. and Spalding, D. B. (1974) 'The numerical computation of fluid flows', *Turbulent Comput. Methods Appl. Mech. Eng.*, **3**, 269–89.
13. Szekely, J., Carlsson, G. and Helle, L. (1989) *Ladle Metallurgy*, Springer: Berlin.

14. Szekely, J. and Ilegbusi, O. J. (1989) *The Physical and Mathematical Modeling of Tundish Operations*, Springer: Berlin.
15. Szekely, J., Evans, J. W. and Brimacombe, J. K. (1988) *The Mathematical and Physical Modeling of Primary Metals Processing Operations*, Wiley: New York.
16. Szekely, J. and Themelis, N. J. (1971) *Rate Phenomena in Process Metallurgy*, Wiley Interscience: New York.
17. Grassmann, P. (1983) *Physikalische Grundlagen der Verfahrenstechnik*, Sauerländer: Frankfurt.
18. Mori, K. (1988) 'Kinetics of fundamental reactions pertinent to steelmaking process', *Trans. Iron Steel Inst. Jpn*, **28**, 246–61.
19. Mendelson, H. D. (1967) 'The prediction of bubble terminal velocities from wave theory', *Am. Inst. Chem. Eng. J.*, **13**, 250–3.
20. Levich, V. G. (1962) *Physicochemical Hydrodynamics*, Prentice Hall: Englewood Cliffs, NJ.
21. Lapple, C. E. (1950) In *Chemical Engineers Handbook* (ed. J. H. Perry), 3rd edn, McGraw-Hill: New York.
22. Calderbank, P. H. (1959) 'Physical rate processes in industrial fermentation, part II: Mass transfer coefficients in gas–liquid contacting with and without agitation', *Trans. Inst. Chem. Eng.*, **37**, 173–85.
23. Oeters, F. (1989) *Metallurgie der Stahlherstellung*, Verlag Stahleisen: Düsseldorf and Springer: Berlin.
24. Steinmetz, E. and Scheller, P. R. (1987) 'Beitrag zu den Strömungsverhältnissen in einer Spülsteinpfanne', Stahl und Eisen, **107**, 417–25.
25. Sano, M. and Mori, K. (1980) 'Size of bubbles in energetic gas injection into liquid metal', *Trans. Iron Steel Inst. Jpn*, **20**, 675–81.
26. Oeters, F., Pluschkell, W., Steinmetz, E. and Wilhelmi, H. (1988) 'Fluid flow and mixing in secondary metallurgy', *Steel Res.*, **59**, 192–201.
27. Nakanishii, K., Fuji, T. and Szekely, J. (1975) 'Possible relationship between energy dissipation and agitation in steel processing operations', *Ironmaking and Steelmaking*, **2**, 193–7.
28. Chung, S.-H. and Lange, K. W. (1988) 'Convective diffusion and dispersion of additions in steel melts', *Ironmaking and Steelmaking*, **15**, 244–56.
29. Woo, J. S., Szekely, J., Castillejos, A. H. and Brimacombe, J. K. (1990) 'A study on the mathematical modelling of turbulent recirculating flows in gas-stirred ladles', *Metall. Trans.*, **21B**, 269–77.
30. Debroy, T., Mazumdar, A. K. and Spalding, D. B. (1978) 'Numerical prediction of recirculation flows with free convection encountered in gas agitated reactors', *Appl. Math. Modelling*, **2**, 146–50.
31. Szekely, J., El-Kaddah, N. H. and Grevet, J. H. (1980) 'Flow phenomena in argon stirred ladles; room temperature measurements and analysis' Scaninject II, *Proc. 2nd Int. Conf. on Injection Metallurgy, Luleå, Sweden, 12–13 June 1980*, Mefos: Luleå, paper no. 5.
32. Mazumdar, D. and Guthrie, R. I. L. (1985) 'Numerical computation of flow and mixing in ladle metallurgy steelmaking operations', *Appl. Math. Modelling*, **10**, 25–32.
33. Spalding, D. B. (1980) 'Numerical computation of multiphase fluid flow and heat transfer', In *Recent Advances in Numerical Methods in Fluids*, Vols 1, 2 (ed. C. Taylor and K. Morgan), Pineridge Press: Swansea, pp. 139–67.
34. Sahai, Y. and Guthrie, R. I. L. (1982) 'Effective viscosity models of gas stirred ladles', *Metall. Trans.*, **13B**, 125–7.

35. Mazumdar, D. (1989) 'On effective viscosity models of gas-stirred ladle systems', *Metall. Trans.* **20B**, 967–9.
36. Grevet, J. H., Szekely, J. and El-Kaddah, N. (1982) 'An experimental and theoretical study of gas bubble driven circulation systems', *Int. J. Heat Mass Transfer*, **25**, 487–97.
37. Szekely, J., Wang, H. J. and Kiser, K. M. (1976) 'Flow pattern velocity and turbulence energy measurements and predictions in a water model of an argon-stirred ladle', *Metall. Trans.*, **7B**, 287–95.
38. Sahai, Y. and Guthrie, R. I. L. (1982) 'Hydrodynamics of gas stirred melts, Part I: Gas–liquid coupling', *Metall. Trans.*, **13B**, 193–202.
39. Ilegbusi, O. J. and Szekely, J. (1990) 'The modeling of gas-bubble driven circulations systems', *Iron Steel Inst. Jpn Int.*, **30**, 731–9.
40. Sano, M. and Mori, K. (1983) 'Circulating flow model in a molten metal bath with special respect to behavior of bubble swarms and its application to gas injection processes', *Scaninject III, Proc. 3rd Int. Conf. on Injection Metallurgy, Luleå, Sweden, 15–17 June 1983*, Mefos: Luleå, paper no. 6.
41. Krishna Murthy, G. G., Ghosh, A. and Mehrotra, S. P. (1989) 'Mathematical modeling of mixing phenomena in a gas stirred liquid bath', *Metall. Trans.*, **20B**, 53–9.
42. Schneider, S. (1988) 'Investigations on energy and mass transfer in gas stirred melts', *Thesis*, Technical University Berlin.
43. Steinmetz, E., Wilhelmi, H., Wimmer, W. and Imo, J. (1983) 'Mischungsvorgänge in Rührreaktoren', *Arch. Eisenhüttenwes.*, **54**, 19–22.
44. Maruyama, T., Kamishima, N. and Mitushima, T. (1984) 'An investigation of bubble plume mixing by comparison with liquid jet mixing', *J. Chem. Eng. Jpn*, **17**, (2), 120–6.
45. Schwarz, M. P. and Turner, W. J. (1988) 'Applicability of the standard k–epsilon turbulence model to gas-stirred baths', *Appl. Math. Modelling*, **12**, 273–9.
46. Krishna Murthy, G. G. (1989) 'Mathematical modelling and simulation of recirculatory flow as well as mixing phenomena in gas stirred liquid baths', *Iron Steel Inst. Jpn Int.*, **29**, 49–57.
47. Mietz, J. and Oeters, F. (1987) 'Mixing theories for gas-stirred melts', *Steel Res.*, **58**, 446–53.
48. Chung, S.-H. and Lange, K. W. (1989) 'Turbulence characteristics of a gas-stirred steel bath outside the bubble plume', *Steel Res.*, **60**, 49–59.
49. Mietz, J. and Oeters, F. (1989) 'Flow field and mixing with eccentric gas stirring', *Steel Res.*, **60**, 387–94.
50. Asai, S., Kawachi, M. and Muchi, I. (1983) 'Mass transfer rate in ladle refining processes', *Scaninject III, Proc. 3rd Int. Conf. on Injection Metallurgy, Luleå, Sweden, 15–17 June 1983*, Mefos: Luleå, paper no. 12.
51. Ebneth, G. and Pluschkell, W. (1985) 'Dimensional analysis of the vertical heterogeneous buoyant plume', *Steel Res.*, **56**, 513–18.
52. Iguchi, M., Tani, J., Uemura, T., Kawabata, H., Takeuchi, H. and Morita, Z.-I. (1989) 'The characteristics of water and bubbling jets in a cylindrical vessel with bottom blowing', *Iron Steel Inst. Jpn Int.*, **29**, 309–17.
53. Koria, S. C. and Singh, S. (1989) 'Measurements on the local properties of the vertical heterogeneous buoyant plume', *Steel Res.*, **60**, 301–7.
54. Koria, S. C. and Singh, S. (1990) 'Studies on the correlationship between structure and physical properties in the buoyant plume rising in liquid', *Steel Res.*, **61**, 287–94.
55. Anagbo, P. E. and Brimacombe, J. K. (1990) 'Plume characteristics and liquid circulation in gas injection through a porous plug', *Metall. Trans.*, **21B**, 637–48.

56. Koria, S. C. (1988) 'Nozzle design in impinging jet steelmaking processes', *Steel Res.*, **59**, 104–9.
57. Mazumdar, D. (1990) 'Dynamic similarity considerations in gas-stirred ladle systems', *Metall. Trans.*, **21B**, 925–8.
58. Johnstone, R. E. and Thring, M. W. (1957) *Pilot Plants, Models and Scale-up Methods in Chemical Engineering*, McGraw-Hill: New York.
59. Kim, S.-H. and Fruehan, R. J. (1987) 'Physical modeling of liquid–liquid mass transfer in gas-stirred ladles', *Metall. Trans.*, **18B**, 381–90.
60. Deo, B. and Mishra, R. K. (1988) 'Cold model study of the effect of blowing regime on refractory wear in topblown converters', *Trans. Ind. Inst. Met.*, **41**, 57–63.
61. Guthrie, R. I. L. (1989) *Engineering in Process Metallurgy*, Clarendon Press: Oxford.

2 *Thermodynamic fundamentals*

2.1 Definition of a system

Thermodynamics is traditionally known as the science of equilibrium. In order to carry out a thermodynamic study, we can choose a small portion of the universe and define it as a system. The system can be enclosed in a boundary and everything in the universe, outside the boundary, is then termed as surroundings. Depending upon the permeability of the boundary, the system can be open, closed or isolated. An open system permits transfer of both matter and energy; a closed system allows the transfer of energy only; an isolated system will allow the transfer neither of matter nor of energy. A system may be further divided into one or more smaller regions. If each region is spatially uniform in characteristics like composition, structure and density, it is called a homogeneous system, consisting of a single or homogeneous phase. Otherwise, it is called a heterogeneous system, consisting of two or more phases. An example of an open, heterogeneous metallurgical system in steelmaking is the oxygen steelmaking converter, as shown in Figure 2.1. The converter is enclosed in a system boundary as indicated by the dashed line. The transfer of mass and energy associated with the charging of scrap, hot metal, ore, oxygen, lance cooling water, flux, etc., and the transfer associated with the output of steel, slag, exhaust gases, etc., are assumed to take place across the fictitious boundary.

2.2 Enthalpy, entropy and free energy changes in a system

The change in enthalpy (ΔH) is represented with respect to a reference state or standard state. The standard state is usually the pure substance (solid, liquid or gas) at 298.14 K and 1 bar pressure.

Consider a system containing m phases. If there are no interactions between phases and if no crystallographic changes occur within the phases in the temperature range ($T_1 - T_2$) of interest, then the net enthalpy increment due to an increase in

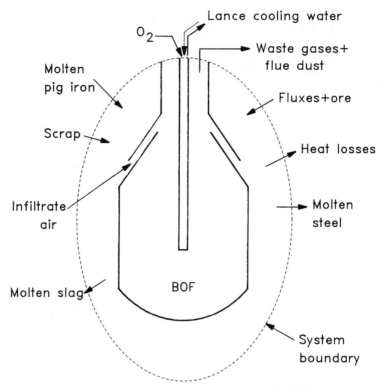

Figure 2.1   A typical open system in steelmaking.

temperature from T_1 to T_2 at constant pressure can be obtained from the summation

$$H^o_{T_2} - H^o_{T_1} = \Delta H = \sum_{i=1}^{m} n_i \int_{T_1}^{T_2} C^i_p \, dT \qquad (2.1)$$

where n_i is the number of moles of atoms of the *i*th phase and C^i_p is the molar heat capacity of the *i*th phase at constant pressure. Over a limited temperature range the heat capacity data are reported in the form of a best-fit equation of the type

$$C_p = a + bT + cT^{-2} \qquad (2.2)$$

where *a*, *b* and *c* are constants.

The heat required to change the crystal structure or the state of aggregation of 1 mole of material is called the heat of transformation, ΔH^{trans}. For a single phase, all heats of transformation falling in the temperature range $(T_1 - T_2)$ of interest are added up to calculate the net enthalpy increment:

$$H^o_{T_2} - H^o_{T_1} = \Delta H = \int_{T_1}^{T_{\text{trans}}} C^1_p \, dT + \Delta H^{\text{trans}} + \int_{T_{\text{trans}}}^{T_2} C^2_p \, dT \qquad (2.3)$$

where C_p^1 and C_p^2 are the heat capacities in the temperature range $T_1 - T_{\text{trans}}$ and $T_{\text{trans}} - T_2$, respectively. There may be more than one phase change in a given temperature range and then the calculation is done for each phase separately; this procedure will become evident from the worked example given below.

Example 2.1 Enthalpy change in heating 1 mole of iron from 298 K to 1873 K, including the crystallographic and phase changes

The phase transformations taking place in iron, in the temperature range 298–1873 K, are as follows:

Reaction	Temperature (K)	Heat effect (J/mol)	Notation
$Fe_\alpha \rightarrow Fe_\beta$	1033	+5105	ΔH_1
$Fe_\beta \rightarrow Fe_\gamma$	1187	+670	ΔH_2
$Fe_\gamma \rightarrow Fe_\delta$	1665	+837	ΔH_3
$Fe_\delta \rightarrow Fe_{\text{liq}}$	1809	+13807	ΔH_4

The C_p values of the different phases [1] are

$$C_{p_\alpha} = 17.49 + 24.769 \times 10^{-3}T \quad \text{J/mol K}$$

$$C_{p_\beta} = 37.66 \quad \text{J/mol K}$$

$$C_{p_\gamma} = 7.70 + 19.5 \times 10^{-3}T \quad \text{J/mol K}$$

$$C_{p_\delta} = 28.284 + 7.531 \times 10^{-3}T \quad \text{J/mol K}$$

$$C_{p_{\text{liq}}} = 35.4 + 3.745 \times 10^{-3}T \quad \text{J/mol K}$$

The enthalpy changes of α, β, γ, δ and liquid iron phases, denoted as ΔH_5, ΔH_6, ΔH_7, ΔH_8 and ΔH_9, respectively, are obtained from the heat capacity data:

$$\Delta H_5 = \int_{298}^{1033} C_{p_\alpha}\, dT = \int_{298}^{1033} (17.49 + 24.769 \times 10^{-3}T)\, dT$$

$$= 24971 \text{ J/mol}$$

$$\Delta H_6 = \int_{1033}^{1187} C_{p_\beta}\, dT = \int_{1033}^{1187} 37.66\, dT$$

$$= 5800 \text{ J/mol}$$

$$\Delta H_7 = \int_{1187}^{1665} C_{p_\gamma}\, dT = \int_{1187}^{1665} (7.70 + 19.5 \times 10^{-3}T)\, dT$$

$$= 16972 \text{ J/mol}$$

$$\Delta H_8 = \int_{1665}^{1809} C_{p_\delta} \, dT = \int_{1665}^{1809} (28.284 + 7.531 \times 10^{-3} T) \, dT$$

$$= 5957 \text{ J/mol}$$

$$\Delta H_9 = \int_{1809}^{1873} C_{p_{\text{liq}}} \, dT = \int_{1809}^{1873} (35.4 + 3.745 \times 10^{-3} T) \, dT$$

$$= 2707 \text{ J/mol}$$

The net enthalpy increment in heating iron from 298 K to 1873 K is the sum of all the enthalpy increments:

$$[\Delta H_{\text{Fe}}]_{298}^{1873} = \sum_{i=1}^{i=9} \Delta H_i$$

$$= 76\,826 \text{ J/mol}$$

In a reacting system, if the products (i) and the reactants (j) are considered to be at the same temperature and pressure, the net heat of reaction is given by Hess' law. Let $\Delta H_{298}^{\circ,\text{react}}$ be the heat of reaction when the reactants and the products are in their standard states at 298 K. At any given temperature T, the expression for the enthalpy change for the system is

$$\Delta H_T^{\circ,\text{react}} = \Delta H_{298}^{\circ,\text{react}} + (H_{T_2}^{\circ} - H_{T_1}^{\circ})_{\text{products}} - (H_{T_2}^{\circ} - H_{T_1}^{\circ})_{\text{reactants}} \tag{2.4}$$

The variation of enthalpy change with temperature is related to the change in heat capacity (ΔC_p) by

$$\left(\frac{d\Delta H}{dT} \right)_p = \Delta C_p \tag{2.5}$$

For the whole system, the change in heat capacity, ΔC_p^{syst}, is given by

$$\Delta C_p^{\text{syst}} = \sum_{\text{products}} n_i (\partial H_i / \partial T)_p - \sum_{\text{reactants}} n_j (\partial H_j / \partial T)_p$$

$$= \sum_{\text{products}} n_i C_{p_i} - \sum_{\text{reactants}} n_j C_{p_j} \tag{2.6}$$

On substituting the expanded form of Equation (2.2):

$$\Delta C_p^{\text{syst}} = \left(\sum_{\text{products}} n_i a_i - \sum_{\text{reactants}} n_j a_j \right) + \left(\sum_{\text{products}} n_i b_i - \sum_{\text{reactants}} n_j b_j \right) T + \left(\sum_{\text{products}} n_i c_i - \sum_{\text{reactants}} n_j c_j \right) T^{-2}$$

Or

$$\Delta C_p^{\text{syst}} = \Delta a + \Delta b T + \Delta c T^{-2} \tag{2.7}$$

Thus, on integrating Equation (2.5) and substituting in Equation (2.4), the enthalpy change for the system is given by

$$\Delta H_T^{\text{o, react}} = \Delta H_{298}^{\text{o, react}} + \int_{298}^{T} \Delta C_p^{\text{syst}} \, dT + \sum_{\text{products}} \Delta H^{\text{trans}} - \sum_{\text{reactants}} \Delta H^{\text{trans}} \tag{2.8}$$

According to the third law of thermodynamics, the entropy of a perfect crystal is zero (J/mol K) at absolute zero temperature. It is thus possible to calculate the absolute value of entropy (unlike that of enthalpy) at 298 K from the heat capacity data:

$$S_{298}^{\text{o}} = \int_{0}^{298} \left(\frac{C_p}{T} \right) dT \tag{2.9}$$

The experimental determination of heat capacity is difficult at low temperatures and it is often necessary to extrapolate the C_p values from higher temperatures to a lower temperature.

Using the thermodynamic relation between Gibbs' free energy (G), enthalpy and entropy (i.e. $G = H - TS$), the standard free energy (Gibbs' free energy) of reaction ($\Delta G_T^{\text{o, react}}$) and standard entropy of reaction ($\Delta S_T^{\text{o, react}}$) are defined in the same way as the standard heat of reaction $\Delta H_T^{\text{o, react}}$. From the knowledge of enthalpy and entropy changes the free energy changes can be calculated from

$$\Delta G_T^{\text{o, react}} = \Delta H_T^{\text{o, react}} - T\Delta S_T^{\text{o, react}} \tag{2.10}$$

2.3 Thermodynamics of liquid iron alloys

Liquid steel is a liquid solution of the atoms of iron (solvent) and the atoms of the alloying elements (solutes) like manganese, carbon, silicon, sulphur, phosphorus, oxygen, hydrogen, nitrogen, chromium, vanadium, titanium, niobium, tungsten, etc. In an ideal solution there are no interactions between different types of atoms and during the formation of the solution the enthalpy and volume changes are zero (i.e. $\Delta H = 0$ and $\Delta V = 0$). The change in entropy is, however, not zero because the addition of new types of atoms to a solvent increases the disorder. It can be shown from statistical thermodynamics that

$$\Delta S = -R \sum_{i} X_i \ln X_i \tag{2.11}$$

where X_i is the mole fraction of the component i. From Equation (2.10)

$$\Delta G = \Delta H - T\Delta S$$

and for an ideal solution $\Delta H = 0$; hence

$$\Delta G = RT \sum_{i} X_i \ln X_i \tag{2.12}$$

Iron and manganese, upon alloying, form a nearly ideal solution. In non-ideal solutions, $\Delta H \neq 0$ and $\Delta V \neq 0$ and the added solute atoms are attracted or repelled, to varying degrees, by the resident atoms in the solution. The intensity of such interactions will decide the departure from ideal behaviour.

According to Raoult's law, in the case of an ideal solution if the activity of a component i (a_i) is plotted as a function of mole fraction (X_i), as shown in Figure 2.2(a), then a straight line will be obtained with a slope of unity, i.e.

$$a_i = X_i \tag{2.13}$$

The dashed line in Figure 2.2(a) is called the Raoult's law line. Most of the solutions are non-ideal and

$$a_i = \gamma_i X_i \tag{2.14}$$

where γ_i is called the Raoultian activity coefficient. It is obvious that only when $\gamma_i = 1$, Raoult's law is obeyed. The deviation from Raoult's law can be positive or negative. For example, the plot in Figure 2.2(b), where a_i is greater than X_i, is a typical example of positive deviation from ideal behaviour (namely, in the Fe–Cu system). Similarly, Figure 2.2(c), where a_i is less than X_i, is an example of negative deviation from ideal behaviour (namely, in the Fe–Si system).

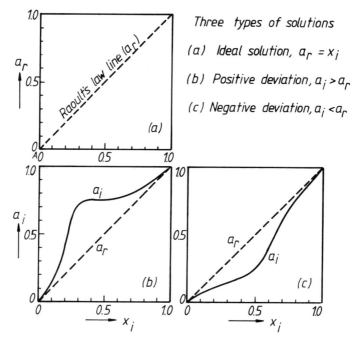

Figure 2.2 Schematic description of: (a) ideal solution; (b) non-ideal solution with positive deviation; and (c) non-ideal solution with negative deviation.

Activity is related to free energy in the following way. The free energy change for a system doing work against pressure p is

$$dG = V\, dp - S\, dT \tag{2.15}$$

where V is the molar volume. At constant temperature

$$dG = V\, dp \tag{2.16}$$

For 1 mole of ideal gas, and since $pV = RT$,

$$dG = RT\, d(\ln p) \tag{2.17}$$

If the gas is non-ideal pressure p is replaced by fugacity f:

$$dG = RT\, d(\ln f) \tag{2.18}$$

Now $a = f/f_o$ and hence

$$dG = RT\, d[\ln(f_o a)] \tag{2.19}$$

Since f_o is constant at a given temperature and composition:

$$dG = RT\, d(\ln a) \tag{2.20}$$

In the standard state, the free energy of a substance is denoted by G_i^o, whereas its partial molar free energy in a solution is denoted by $\bar{G}_i$. Integrating Equation (2.20):

$$\int_{G_i^o}^{\bar{G}_i} dG = \int_{a_o}^{a_i} RT\, d(\ln a) = \bar{G}_i - G_i^o = RT \ln a_i - RT \ln a_o \tag{2.21}$$

where a_o is the activity of the pure substance, i.e. $a_o = 1$ and

$$\Delta \bar{G}_i^M = \bar{G}_i - G_i^o = RT \ln a_i \tag{2.22}$$

$\Delta \bar{G}_i^M$ is the relative partial molar free energy of solution of the component i in the solution. If the solution consists of only two components A and B, the integral molar free energy of mixing, ΔG^M, can be obtained from Equation (2.22) as

$$\Delta G^M = RT(X_A \ln a_A + X_B \ln a_B) \tag{2.23}$$

Equations (2.22) and (2.23) provide a direct link between an experimentally determinable quantity a_i (through the vapour pressure measurements) and the free energy. From Equation (2.15)

$$\Delta \bar{G}_i^M = \Delta \bar{V}_i^M\, dp - \Delta \bar{S}_i^M\, dT$$

Hence

$$\left(\frac{\partial \bar{G}_i}{\partial T} \right)_p = \left(\frac{\partial}{\partial T} (RT \ln a_i) \right)_p = -\Delta \bar{S}_i^M$$

or

$$\Delta \bar{S}_i^M = -R \ln a_i - RT \left(\frac{\partial (\ln a_i)}{\partial T} \right)_p \tag{2.24}$$

Also, by definition,

$$\Delta \bar{G}_i^M = \Delta \bar{H}_i^M - T \Delta \bar{S}_i^M$$

From Equations (2.22) and (2.24)

$$\Delta \bar{H}_i^M = -RT^2 \left(\frac{\partial \ln a_i}{\partial T} \right)_P \tag{2.25}$$

By knowing the variation of activity with temperature, one can calculate $\Delta \bar{S}_i^M$ and $\Delta \bar{H}_i^M$ at a given temperature.

The free energy change during a chemical reaction is also related to the activity and can be calculated as follows.

Let there be a reaction of the following form:

$$aA + bB + \cdots = cC + dD + \cdots \tag{2.26}$$

where a moles of A, b moles of B,... combine to give c moles of C and d moles of D,.... In a general case, where the components are considered to be present in solution, the free energy change of the reaction is given by

$$\Delta G = [c\bar{G}_C + d\bar{G}_D + \cdots] - [a\bar{G}_A + b\bar{G}_B + \cdots] \tag{2.27}$$

Hence from equation (2.22)

$$\Delta G = [c(G_C^\circ + RT \ln a_C) + d(G_D^\circ + RT \ln a_D) + \cdots]$$
$$\qquad - [a(G_A^\circ + RT \ln a_A) + b(G_B^\circ + RT \ln a_B) + \cdots]$$
$$\qquad = [(cG_C^\circ + dG_D^\circ + \cdots) - (aG_A^\circ + bG_B^\circ + \cdots)]$$
$$\qquad + RT \ln \left(\frac{a_C^c a_D^d \cdots}{a_A^a a_B^b \cdots} \right)$$
$$\qquad = \Delta G^\circ + RT \ln Q \tag{2.28}$$

where

$$Q = \frac{a_C^c a_D^d \cdots}{a_A^a a_B^b \cdots} \tag{2.29}$$

ΔG° is the free energy change of the reaction when all the reactants and products are in their standard states. At equilibrium $\Delta G = 0$; hence

$$\Delta G^\circ = -RT \ln(Q)_{\text{equilibrium}} \tag{2.30}$$

Since ΔG° is a constant at a given temperature and pressure, $(Q)_{\text{equilibrium}}$ must also be a constant. This is known as the equilibrium constant K; hence

$$\Delta G^\circ = -RT \ln K \tag{2.31}$$

The equilibrium concentrations of different elements taking part in a chemical reaction in a closed system at constant temperature and pressure can change depending upon their initial concentrations, but only in such a way that K remains constant. This is the basic principle of refining of liquid steel.

The chemical reactions taking place in a closed system also obey the phase rule, according to which

$$F = C - P + 2 \qquad (2.32)$$

where F is degrees of freedom or minimum number of additional state variables required to be fixed for a specified system, C is the number of components in the system, and P is the number of phases. Thus, for a two-phase, two-component system there would be two degrees of freedom. If temperature and pressure are fixed (special restrictions imposed on the system) then the degree of freedom would be zero and the composition of each phase would be fixed.

For a reacting system with i chemical species, r independent chemical reactions, P phases, x stoichiometric relations and n_r special constraints (restrictions already imposed on the system)

$$F = C - P + 2$$

where

$$C = i - r - x - n_r$$

Consider the oxidation of carbon by oxygen gas in a molten Fe–C–O alloy at constant temperature and pressure. Three possible chemical reactions are:

$$[C] + [O] = (CO)_g$$
$$[C] + 2[O] = (CO_2)_g$$
$$[Fe] + [O] = (FeO)$$

- There are three independent chemical reactions: $r = 3$.
- The species are CO, CO_2, C, O, Fe and FeO: $i = 6$.
- The reacting phases are slag, metal, gas: $P = 3$.
- Special constraints are constant temperature and pressure: $n_r = 2$.
- There is no stoichiometric constraint: $x = 0$.

Thus $F = 6 - 3 - 0 - 2 - 3 + 2 = 0$, i.e. the system has zero degrees of freedom.

2.3.1   *Dilute solutions in liquid steel*

Elements like oxygen, carbon, hydrogen, nitrogen, silicon, manganese, sulphur and phosphorus are generally present in small concentrations in liquid steel, i.e. in the form of a dilute solution. The thermodynamic treatment of dilute liquid steel solutions is greatly simplified by Henry's law, according to which the Henrian activity of any component at infinite dilution is equal to its concentration. During day-to-day operations, the operator in a steel plant prefers to express the concentration in wt % rather than in mole % (or mole fractions). According to Henry's law for dilute solutions, the Henrian activity (h_i) of a solute i is equal to the concentration expressed in wt %:

$$h_i = (\text{wt \%})_i \qquad\qquad (2.33)$$

The term 'dilute' may be used as long as the above relationship is valid and thus the range of application of Equation (2.33) may vary from one element to another. From thermodynamics, however, it can be shown that Equation (2.33) is valid only at infinite dilutions. If the concentration is expressed in mole fractions X_i, then, according to Henry's law for a solution at infinite dilution,

$$h_i' = X_i \qquad\qquad (2.34)$$

Both h_i and h_i' are related to Raoult's law. In order to derive the relationship between the Raoultian and the Henrian activities, it is necessary to understand the concept of standard states.

   The solid line in Figure 2.3 depicts a marked negative deviation from ideal behaviour and the dashed line AB represents the Raoult's law line. Up to point C (say 0.3 mole fractions), the Raoultian activity a_i may be assumed to fall on the straight line ACD. The line ACD also represents Henry's law line corresponding to Equations (2.33) and (2.34). The ratio FD/AF gives the slope of line ACD and if this is denoted by γ_i°, then in the region AC the Raoultian activity a_i will be given by

$$a_i = \gamma_i^\circ X_i \qquad\qquad (2.35)$$

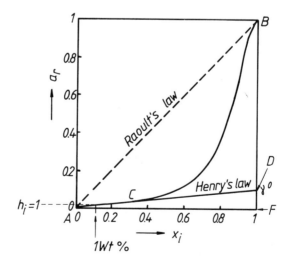

Figure 2.3   Relationship between Raoult's law and Henry's law.

The numerical relationship between h_i, h_i' and a_i, on the basis of Equations (2.33)–(2.35), may now be written as

$$\frac{h_i}{(\text{wt \%})_i} = \frac{h_i'}{X_i} = \frac{a_i}{\gamma_i^\circ X_i} = 1 \tag{2.36}$$

It follows that

$$a_i = \gamma_i^\circ X_i \frac{h_i}{(\text{wt \%})_i} \tag{2.37}$$

and

$$a_i = h_i' \gamma_i^\circ \tag{2.38}$$

From Equation (2.22), the relative partial molar free energy of a substance in solution is related to the Raoultian activity, a_i, by

$$\Delta \bar{G}_i^M = RT \ln a_i$$

Three cases of a_i may now be considered:

Case I $\quad a_i = 1$ (pure substance)

Hence

$$\Delta \bar{G}_i^M = RT \ln 1 = 0 \tag{2.39}$$

Case II $\quad a_i = \gamma_i^\circ X_i h_i / (\text{wt \%})_i$ (from Equation (2.37))

When $h_i = 1$, $(\text{wt \%})_i = 1$ (from Equation (2.33)). It can also be shown, as a first approximation, that at infinite dilution in iron the mole fraction of the solute i is related to the wt % of i by

$$X_i = \frac{55.85}{100 M_i} (\text{wt \%})_i \tag{2.40}$$

where 55.85 is the atomic weight of iron and M_i is the atomic weight of the element i. The relative partial free energy of solution is therefore

$$\Delta \bar{G}_i^M = RT \ln a_i = RT \ln [\gamma_i^\circ (0.5585 / M_i)(\text{wt \%})_i] \tag{2.41}$$

The free energy change $\Delta \bar{G}_1^\circ$ associated with the change of standard state from pure substance to 1 wt %, i.e.

$$i_{\text{pure}} \rightarrow [i]_{1 \text{ wt \%}}$$

is thus given by

$$\Delta \bar{G}_i^\circ = RT \ln [\gamma_i^\circ (0.5585 / M_i)]$$

Case III $a_i = h_i' \gamma_i^\circ$ (from Equation (2.38))

When $h_i' = 1$,

$$\Delta \bar{G}_i^M = \Delta \bar{G}_2^\circ = RT \ln a_i = RT \ln \gamma_i^\circ \tag{2.42}$$

$\Delta \bar{G}_2^\circ$ is the free energy change due to the change of standard state from pure substance to infinite dilution.

Beyond point C in Figure 2.3, Henry's law is not obeyed and Equation (2.36) becomes

$$\frac{h_i}{(\mathrm{wt}\,\%)_i} = \frac{h_i'}{X_i} = \frac{a_i}{\gamma_i^\circ X_i} = f_i \tag{2.43}$$

where the term f_i, known as the Henrian activity coefficient, accounts for the deviation from Henry's law.

The equations (2.41) and (2.42) now become

$$\Delta \bar{G}_1^\circ = RT \ln \left(f_i \gamma_i^\circ X_i \right) \tag{2.44}$$

$$\Delta \bar{G}_2^\circ = RT \ln(f_i \gamma_i^\circ) \tag{2.45}$$

For most of the solutes dissolved in liquid iron in the form of a dilute solution, the term $\ln f_i$ varies linearly with mole fraction of the solute or $\lg f_i$ varies linearly with wt % of the solute. From a Taylor's series expansion one can write the first term as

$$\ln f_i = \frac{\partial (\ln f_i)}{\partial X_i} X_i \tag{2.46}$$

and

$$\lg f_i = \frac{\partial (\lg f_i)}{\partial (\mathrm{wt}\,\%)_i} (\mathrm{wt}\,\%)_i \tag{2.47}$$

The term $\partial(\ln f_i)/\partial X_i$ is called the interaction coefficient ε_i^i and the term $\partial(\lg f_i)/\partial(\mathrm{wt}\,\%)_i$, the interaction coefficient e_i^i. At high dilution the relationship between e_i^i and ε_i^i can be written from Equations (2.46) and (2.47) as

$$2.303 e_i^i (\mathrm{wt}\,\%)_i = \varepsilon_i^i X_i \tag{2.48}$$

As shown in Equation (2.40), for liquid iron at high dilution

$$X_i = \frac{0.5585}{M_i} (\mathrm{wt}\,\%)_i$$

Hence

$$2.303 e_i^i (\mathrm{wt}\,\%)_i = \varepsilon_i^i \frac{0.5585}{M_i} (\mathrm{wt}\,\%)_i$$

or

$$e_i^i = \frac{0.2425}{M_i} \varepsilon_i^i \tag{2.49}$$

If the solutes $i, j, k, \dots$ are present in dilute solution, i.e. in multicomponent systems, Equations (2.46) and (2.47) can be written in the form of a Taylor's series expansion, neglecting the second- and higher-order terms, as

$$\ln f_i = \frac{\partial(\ln f_i)}{\partial X_i} X_i + \frac{\partial(\ln f_i)}{\partial X_j} X_j + \frac{\partial(\ln f_i)}{\partial X_k} X_k + \cdots$$

$$= \varepsilon_i^i \times X_i + \varepsilon_i^j \times X_j + \varepsilon_i^k \times X_k + \cdots \tag{2.50}$$

and

$$\lg f_i = \frac{\partial(\lg f_i)}{\partial(\text{wt \%})_i} \times (\text{wt \%})_i + \frac{\partial(\lg f_i)}{\partial(\text{wt \%})_j} \times (\text{wt \%})_j$$

$$+ \frac{\partial(\lg f_i)}{\partial(\text{wt \%})_k} \times (\text{wt \%})_k + \cdots \tag{2.51}$$

or

$$\lg f_i = e_i^i(\text{wt \%})_i + e_i^j(\text{wt \%})_j + e_i^k(\text{wt \%})_k + \cdots \tag{2.52}$$

In general, e_i^k and ε_i^k represent the effect of element k on the Henrian activity coefficient of element i, when both i and k are present in dilute solution. The equilibrium constants for the reactions involving the elements dissolved in liquid steel are summarized in Table 2.1. The interaction parameter values (e_i^k) are summarized in Table 2.2.

In some cases, when the variation of $\lg f_i$ (or $\ln f_i$) against concentration is not linear, the second-order terms can be incorporated into Equations (2.51) and (2.52) and these give rise to second-order interaction coefficients. In the treatment of dilute solutions in liquid iron, it generally suffices to use first-order interaction coefficients only.

From Equation (2.43), the Raoultian activity a_i is given by

$$a_i = f_i \gamma_i^\circ X_i \tag{2.53}$$

Since by definition

$$f_i \gamma_i^\circ = \gamma_i \tag{2.54}$$

then

$$a_i = \gamma_i X_i \tag{2.55}$$

where γ_i is the same as the Raoultian activity coefficient appearing in Equation (2.14).

Table 2.1 Equilibrium constants for reactions involving elements dissolved in liquid iron

Reaction	lg K	K at 1600 °C
$(FeO) = \{Fe\} + [O]$	$-(6372/T) + 2.73$	0.212
Oxides:		
$FeAl_2O_4 = \{Fe\} + 2[Al] + 4[O]$	$-(70\,320/T) + 23.38$	6.9×10^{-15}
$MnAl_2O_4 = [Mn] + 2[Al] + 4[O]$	*	0.9×10^{-15}
$Al_2O_3 = 2[Al] + 3[O]$ (up to 0.4 wt % Al)	$-(58\,473/T) + 17.74$	3.3×10^{-14}
$B_2O_3 = 2[B] + 3[O]$	*	1.5×10^{-8}
$CaO = [Ca] + [O]$	*	6.2×10^{-11}
$Ce_2O_3 = 2[Ce] + 3[O]$	$-(68\,500/T) + 19.60$	1.0×10^{-17}
$FeCr_2O_4 = \{Fe\} + 2[Cr] + 4[O]$		
(wt % Cr < 0.3)	$-(45\,796/T) + 18.83$	2.4×10^{-6}
$Cr_2O_3 = 2[Cr] + 3[O]$ (wt% Cr from 3–8%)	$-(45\,531/T) + 20.25$	1.1×10^{-4}
$HfO_2 = [Hf] + 2[O]$	*	8.7×10^{-12}
$La_2O_3 = 2[La] + 3[O]$	$-(62\,050/T) + 14.10$	9.3×10^{-20}
$MgO = [Mg] + [O]$	*	1.0×10^{-6}
$(FeO-MnO)_1 = \{Fe\} + [Mn] + 2[O]$	$-(13\,430/T) + 6.10$	8.5×10^{-2}
$(FeO-MnO)_s = \{Fe\} + [Mn] + 2[O]$	$-(15\,050/T) + 6.73$	5.0×10^{-2}
$FeNb_2O_6 = \{Fe\} + 2[Nb] + 6[O]$	$-(88\,300/T) + 36.76$	4.1×10^{-11}
$NbO_2 = [Nb] + 2[O]$	$-(32\,780/T) + 13.92$	2.6×10^{-4}
$SiO_2 = [Si] + 2[O]$ (<3 wt % Si)	$-(31\,040/T) + 12.00$	2.7×10^{-5}
$Ca_2SiO_4 = 2[Ca] + [Si] + 4[O]$	$-(37\,950/T) + 12.52$	1.8×10^{-8}
$FeTa_2O_6 = \{Fe\} + 2[Ta] + 6[O]$ (<0.25 wt % Ta)	$-(79\,300/T) + 28.43$	1.2×10^{-14}
$Ta_2O_5 = 2[Ta] + 5[O]$	$-(63\,100/T) + 21.90$	1.6×10^{-12}
$TiO_2 = [Ti] + 2[O]$	*	5.0×10^{-7}
$Ti_3O_5 = 3[Ti] + 5[O]$	$-(91\,304/T) + 29.34$	3.9×10^{-20}
$Ti_2O_3 = 2[Ti] + 3[O]$	*	2.7×10^{-11}
$FeV_2O_4 = \{Fe\} + 2[V] + 4[O]$	$-(46\,452/T) + 17.42$	4.2×10^{-8}
$V_2O_3 = 2[V] + 3[O]$	*	3.5×10^{-6}
$VO = [V] + [O]$	$-(15\,530/T) + 6.66$	2.3×10^{-2}
$ZrO_2 = [Zr] + 2[O]$	$-(41\,258/T) + 11.86$	6.8×10^{-11}
Sulphides:		
$CaS = [Ca] + [S]$	*	1.3×10^{-9}
$CeS = [Ce] + [S]$	$-(20\,600/T) + 6.39$	2.5×10^{-5}
$LaS = [La] + [S]$	$-(26\,000/T) + 8.98$	1.3×10^{-5}
$MgS = [Mg] + [S]$	*	3.0×10^{-3}
$MnS = [Mn] + [S]$	*	2.7
$TiS = [Ti] + [S]$	$-(8000/T) + 4.02$	0.56
$ZrS = [Zr] + [S]$	*	0.3
Equilibria involving gases ($p = 1$ bar):		
$\frac{1}{2}(H_2)_g = [H]$	$-(1670/T) - 1.68$	2.68×10^{-3}
$\frac{1}{2}(N_2)_g = [N]$	$-(255.6/T) - 1.26$	4.01×10^{-2}
$\frac{1}{2}(O_2)_g = [O]$	$(6126/T) + 0.15$	2.64×10^3
$(CO)_g = [C] + [O]$	$-(2423.26/T) - 1.367$	2.18×10^{-3}
$(CO_2)_g = (CO)_g + [O]$	$-(10\,415.23/T) + 5.623$	1.154
$(H_2O)_g = 2[H] + [O]$	$-(10\,610/T) - 0.99$	2.21×10^{-7}

Note: For the reactions marked *, reliable values are available at 1600 °C only.

Sources:

(a) Most of the values are adapted from: Olette, M. and Gatellier, C., *Information Symposium on Casting and Solidification of Steel, Luxembourg, 29 November–1 December 1977*, IPC Science and Technology: Guildford, paper 1, pp. 8–61.

(b) Fe–Cr–O, Fe–Al–O, Fe–Zr–O, Fe–V–O and Fe–Ti–O are taken from Ghosh and Murthy [50].

(c) CO–C–O and CO_2–CO–O values taken from Deo [49].

The Henrian activity h_i is obtained from Equation (2.43) as

$$h_i = (\text{wt \%})_i f_i \tag{2.56}$$

or

$$\lg h_i = \lg(\text{wt \%})_i + \lg f_i \tag{2.57}$$

In a multicomponent system $\lg f_i$ is obtained from Equation (2.52).

2.3.2 Thermodynamics of oxygen dissolution in liquid iron

The dissolution of oxygen in liquid steel may be represented by the equation

$$\tfrac{1}{2}(O_2)_g = [O] \tag{2.58}$$

where $[O]$ denotes the oxygen dissolved in the metal as atomic oxygen with a reference state of 1 wt %. The equilibrium constant K_O for Equation (2.58) can be written as

$$K_O = \frac{h_O}{(p_{O_2})^{1/2}} \tag{2.59}$$

The standard free energy is given by

$$\Delta G_O^\circ = -RT \ln K_O$$
$$= -117.3 - 2.889 \times 10^{-3} T \quad \text{kJ/mol} \tag{2.60}$$

where T is the temperature in kelvin, R is the universal gas constant (equal to 8.314 J/mol K), p_{O_2} denotes the partial pressure of oxygen in the gas phase in atmospheres, and h_O is the activity of dissolved oxygen in steel in the 1 wt % standard state. Combining Equations (2.59) and (2.60) yields

$$\lg h_O - \tfrac{1}{2}\lg p_{O_2} = (6126/T) + 0.15 \tag{2.61}$$

Using Equation (2.56)

$$h_O = f_O(\text{wt \%})_O \tag{2.62}$$

where f_O designates the activity coefficient of dissolved oxygen in steel. In pure liquid iron containing dissolved oxygen, the only interaction parameter to be considered is $e_O^O = -0.20$. Thus from Equation (2.52):

$$\lg f_O = -0.20[\text{wt \% O}] \tag{2.63}$$

The above relations can be used to estimate the equilibrium value of oxygen [wt % O] in liquid iron at a given temperature and for a given value of p_{O_2}. When liquid iron becomes saturated with oxygen, iron oxide forms, i.e. liquid iron and iron oxide are in equilibrium (see the top left-hand region in the Fe–O phase equilibrium diagram in Figure 2.4). The iron oxide, in equilibrium with pure iron, is denoted as Fe_xO, where x is approximately 0.985 at 1873 K. For the sake of simplicity, x can

Table 2.2 Interaction coefficients of elements dissolved in liquid iron at 1873 K.

PART 1:   Element j from Al to N

i \ $j=$	Al	B	C	C$_{saturated}$	Ce	Cr	H	La	Mn	Mo	N
Al	0.043		0.091	−0.019			0.24				−0.58
B		0.038	0.22	0.009			0.49				0.074
C	0.042	0.24	0.14			−0.024	0.67		−0.012	−0.008	0.11
C$_{saturated}$	0.0069	0.021				−0.015			−0.007	−0.007	0.13
Ce				−0.139				−0.027			
Cr	0.013	0.05	−0.0118			−0.0003	0.60			0.002	−0.19
H			0.06			−0.0022	−0.33		−0.0014	0.002	
La					0.0		0.0				
Mn				−0.108			−4.3		−0.003		−0.091
Mo	−0.028	0.094	−0.097	−0.192		−0.0003	−0.31				−0.10
N			0.130	0.18		−0.047	−0.20		−0.020	−0.011	0
Nb			−0.49	0.276		−0.0003	−0.61				−0.42
Ni			0.042	−0.018			−0.25				0.028
O	−2.81	−0.31	−0.35	−0.44	−0.57	−0.037	3.1	−0.57	−0.03	0.0035	0.057
P			0.13	−0.005		0.03	0.21		0		0.094
S	0.035	0.13	0.113	0.003		0.010	0.120		−0.025	0.003	0.01
Si	0.058	0.20	0.18	0.177		0.0003	0.64		0.033		0.09
Ta				−0.363			−4.4				−0.47
Ti				−0.229		0.055	−1.1				−1.8
U											
V											
Zr			−0.34	−0.181			−0.72				−0.35

PART 2: Element j from Nb to Zr

i \ $j =$	Nb	Ni	O	P	S	Si	Ta	Ti	U	V	Zr
Al			−4.77		−0.030	0.056			0.011		
B			−0.21		0.047	0.078					
C	−0.060	0.012	−0.26	0.051	0.045	0.08				−0.077	
C$_{saturated}$	−0.018	0.012	−0.22	0.013			−0.02	−0.039			
Ce			−5								
Cr	−0.0023	−0.0002	−0.12	−0.053	−0.02	−0.004		0.06			
H		0	−0.19	0.011	0.008	0.026	−0.02	−0.02		−0.007	
La			−5								
Mn			−0.10	−0.0035	−0.048	0.060					
Mo			−0.0007		−0.0005						
N	−0.06	0.010	−0.05	0.045	0.007	0.045	−0.032	−0.53		−0.093	−0.63
Nb	0.006		−0.7		−0.046						
Ni		0.0009	0.01	−0.0035	−0.004	0.006					
O	−0.12	0.006	−0.20	−0.07	−0.133	−0.14	−0.11	−0.42	−0.44		−3.41
P		0	+0.13	0.62	0.028	0.12					
S		0	0.27	0.29	−0.028	0.065		−0.07			−0.053
Si	−0.013	0.005	−0.25	0.11	0.056	0.11				0.16	
Ta			−1.3		−0.02		0.002				
Ti			−1.27		−0.11			0.056			
U			−6.6						0.01		
V					−0.028	0.27				0.015	
Zr			−19.5		−0.16						0.023

Sources:

(a) Most of the interaction parameter values appearing in this table have been adapted from: Olette, M. and Gatellier, C., *Information Symposium on Casting and Solidification of Steel, Luxemburg, 29 November–1 December 1977*, IPC Science and Technology: Guildford, paper 1, pp. 8–61, and from: Sigworth, G. K. and Elliot, J. F. (1974) 'The thermodynamics of liquid dilute iron alloys', *Met. Sci.*, **8**, 298–310.

(b) e_C^C and e_O^C values are taken from Deo [49].

(c) e_{Al}^O, e_O^{Al}, e_{Zr}^O, e_O^{Zr}, e_{Ti}^O and e_O^{Ti} values are taken from Ghosh and Murthy [50].

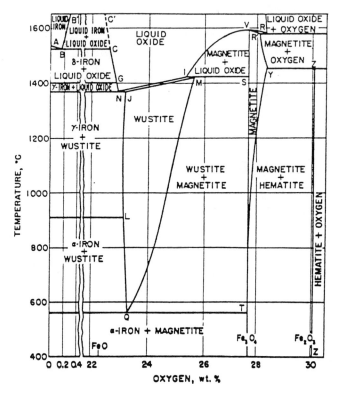

Figure 2.4   Fe–O phase equilibrium diagram. (*Source*: Darken, L. S. and Gurry, R. W., *J. Am. Chem. Soc.*, **67** (1945) 1398–412; **68** (1946) 798–816.)

be assumed to be unity. The general equation for calculating the solubility of oxygen in liquid iron, in equilibrium with FeO, is

$$(FeO) = \{Fe\} + [O] \tag{2.64}$$

The equilibrium constant for the above reaction can be written as

$$K_{Fe} = \frac{[h_O][a_{Fe}]}{(a_{FeO})} \tag{2.65}$$

and

$$\lg K_{Fe} = -(6372/T) + 2.73 \tag{2.66}$$

where a_{Fe}, the activity of iron in the metal phase in the Raoultian scale, is approximately equal to 1, and a_{FeO} denotes the activity of FeO. For $a_{FeO} = 1$ (i.e. if FeO is a pure substance), the maximum solubility of oxygen, from Equation (2.65), is given by

$$K_{Fe} = [h_O]_{saturated} \tag{2.67}$$

If FeO is impure, for instance when it is present in a slag phase along with other components, then $a_{FeO} < 1$, and h_O, i.e. the activity of dissolved oxygen in metal in equilibrium with the slag, will be lower. Using Equations (2.67) and (2.65)

$$K_{Fe} = [h_O]_{saturated} = h_O/a_{FeO} \qquad (2.68)$$

The Henrian activity of oxygen in liquid steel at a given temperature is thus directly dependent upon the activity of iron oxide in the slag, or in other words, the oxidizing power of the slag.

***Example 2.2* Calculate the maximum solubility of oxygen in pure iron at 1873 K**

The Henrian activity of oxygen is related to its concentration in wt % by

$$lg [h_O]_{saturated} = lg f_O + lg [wt \%O]$$

On substituting the value of $lg f_O$ from Equation (2.63)

$$lg [h_O]_{saturated} = -0.20[wt \%O] + lg [wt \%O]$$

Again, at 1873 K, from Equations (2.66) and (2.67)

$$lg K_{Fe} = -0.672 = lg [h_O]_{saturated}$$

Writing $[h_O]_{saturated}$ in terms of $[wt \%O]$

$$-0.20[wt \%O] + lg [wt \%O] = -0.672$$

The above non-linear equation can be solved by a suitable numerical procedure (e.g. the Newton–Raphson method) to yield

$$[wt \%O] = 0.235 \, wt \%$$

Note: $[wt \%O]$ at 1723 K and 1923 K are 0.185 and 0.29, respectively.

2.3.3 *Thermodynamics of nitrogen and hydrogen dissolution in liquid iron*

The dissolution of nitrogen in iron is represented by

$$\tfrac{1}{2}(N_2)_g = [N] \qquad (2.69)$$

According to Sievert's law

$$[\%N] = K_N \sqrt{p_{N_2}} \qquad (2.70)$$

The equilibrium constant for the above reaction is

$$lg K_N = -(255.6/T) - 1.26 \qquad (2.71)$$

Thus, at equilibrium the percentage of nitrogen dissolved in liquid steel is directly proportional to the square root of its partial pressure in the gas phase. Elements such as aluminium, vanadium, titanium and niobium form stable nitrides. If the nitrides are unable to float out during the solidification of steel then the total nitrogen content (i.e. dissolved nitrogen plus nitrogen present as nitrides of aluminium, titanium, etc.) of steel increases. The physical properties of steel are greatly affected by the precipitation of nitrides; for example, aluminum nitride (AlN) causes grain refinement.

The equation for the solubility of hydrogen in liquid iron is

$$\tfrac{1}{2}(H_2)_g = [H] \tag{2.72}$$

$$[H] = K_H\sqrt{p_{H_2}} \tag{2.73}$$

The equilibrium constant (when $[H]$ is in ppm) is

$$\lg K_H = -(1670/T) - 1.68 \tag{2.74}$$

Titanium and tantalum increase the solubility of hydrogen in liquid iron. Unlike nitrogen, the hydrogen in liquid steel is absorbed primarily from the water vapour in the ambient atmosphere:

$$(H_2O)_g = 2[H] + [O] \tag{2.75}$$

$$\Delta G^\circ = 207\,653 - 0.292T \text{ J/mol } H_2O \tag{2.76}$$

Example 2.3 Precipitation of aluminium nitride

Suppose liquid iron at 1873 K contains 0.2% aluminium and 300 ppm nitrogen. Calculate the temperature at which aluminium nitride will begin to precipitate. Assume all interaction parameters are zero.

From thermodynamic data:

$$\{Al\} + \tfrac{1}{2}(N_2)_g = \langle AlN \rangle \quad \Delta G^\circ = -327\,063 + 115.520T \text{ J/mol}$$

$$[N] \qquad\qquad = \tfrac{1}{2}(N_2)_g \quad \Delta G^\circ = \qquad -3598 - 23.891T \text{ J/mol}$$

$$[Al] \qquad\qquad = \{Al\} \qquad \Delta G^\circ = \qquad 63\,178 + 27.90T \text{ J/mol}$$

$$[Al] + [N] \quad = \langle AlN \rangle \, \Delta G^\circ = -267\,483 + 119.529T \text{ J/mol}$$

Since the interaction parameters are zero the equilibrium constant for the above reaction can be written as

$$K_T = \frac{a_{AlN}}{[\text{wt \% Al}] \times [\text{wt \% N}]}$$

Assuming that AlN is pure solid, $a_{AlN} = 1$ and

$$K_T = \frac{1}{(0.2)(0.0300)}$$

From free-energy data:

$$-RT \ln K_T = -267\,483 + 119.529\,T \text{ J/mol}$$

Substituting the value of K_T, the temperature at which aluminium nitride will precipitate is $T = 1650\,\text{K}$.

2.3.4   *Thermodynamics of carbon dissolution in liquid iron*

Carbon is one of the most widely studied elements in steelmaking. The equation for the solubility of carbon (graphite saturation) in a binary Fe–C alloy is given [2] by

$$[\%\,C]_{\text{saturated}} = 0.56 + 2.57 \times 10^{-3}\,T \tag{2.77}$$

The ratio of Raoultian activity coefficient (γ_C) to the activity coefficient at infinite dilution γ_C^o, is given by [3]:

$$\lg(\gamma_C/\gamma_C^o) = -2.38(X_{Fe}^2 - 1) \tag{2.78}$$

where

$$\gamma_C^o = (1180/T) - 0.8 \tag{2.79}$$

The solubility of carbon in iron is also affected by the presence of other alloying elements like silicon, phosphorus, sulphur and manganese.

2.4   Thermodynamics of steelmaking slags

The thermodynamic treatment of steelmaking slags can be based either on the molecular nature of slag or on the ionic nature of slag.

2.4.1   *Molecular theory of slags*

The liquid slag is assumed to be a solution of simple molecules like CaO, MgO, SiO_2, P_2O_5 and complex molecules formed as a result of the chemical reaction between simple molecules. The partitioning of an element between slag and metal will depend upon the respective activities in the metal and slag phases. For example, if silicon in the metal is oxidized to SiO_2, the reaction taking place is

$$[Si] + 2[O] = (SiO_2) \tag{2.80}$$

The partitioning of silicon between slag and metal will thus depend upon the activities of silicon and oxygen in the metal and on the activity of silica in the slag phase according to the equilibrium constant:

$$K = \frac{a_{SiO_2}}{[h_{Si}][h_O]^2} \tag{2.81}$$

In the early years of development of the molecular theory of slag, complex molecules such as $3CaO \cdot P_2O_5$, $4CaO \cdot P_2O_5$, $CaO \cdot SiO_2$ and $2CaO \cdot SiO_2$ were assumed to be non-dissociated and non-reactive. It was suggested that only the basic oxides like CaO and MgO, present in excess of the requirement to form complex molecules, participated in slag–metal reactions. This, however, imposed the limitation that a slag consisting of only pure $2CaO \cdot SiO_2$ could not have influenced the refining reactions, which was not found to be true. As an improvement, a partial dissociation of the complex molecules was proposed, namely

$$(2CaO \cdot SiO_2) = 2(CaO) + (SiO_2) \tag{2.82}$$

The dissociation constants were determined for such reactions.

Several empirical relationships based on the molecular theory of slag are used in steel plants. For example, the V ratio (or basicity) is defined as

$$V = \frac{(wt \% \, CaO)}{(wt \% \, SiO_2)} \tag{2.83}$$

Modified forms of the V ratios are also employed:

$$\text{Modified } V \text{ ratio} = \frac{(wt \% \, CaO)}{(wt \% \, SiO_2) + (wt \% \, Al_2O_3) + (wt \% \, P_2O_5)} \tag{2.84}$$

$$\text{Basicity } B_1 = \frac{\sum wt \% \, (\text{all basic oxides})}{\sum wt \% \, (\text{all acid oxides})} \tag{2.85}$$

and

$$\text{Basicity } B_2 = \frac{(wt \% \, CaO) + (wt \% \, MgO)}{(wt \% \, SiO_2) + (wt \% \, Al_2O_3) + (wt \% \, P_2O_5)} \tag{2.86}$$

The slags, at the end of refining in basic oxygen steelmaking, have a basicity greater than 2 (see section 2.4.4).

Example 2.4 **Sulphur distribution between slag and metal**

1. What are the conditions for obtaining a high sulphur distribution between slag and metal?
2. Molten iron containing 3.5% carbon, 0.25% manganese, 0.25% phosphorus and 1% silicon is in equilibrium with pure lime at 1583 K. Calculate the equilibrium sulphur content of the metal.

Note: The interaction parameter values are given in Table 2.2 and the free energy data are given in Table 2.1.

Solution

Part 1

In the absence of carbon, silicon and aluminium in the metal, the desulphurization reaction can be written as

$$[S] + [Fe] + (CaO) = (CaS) + (FeO) \tag{a}$$

$$K = \frac{a_{CaS}a_{FeO}}{h_S a_{Fe} a_{CaO}} \tag{b}$$

The mole fraction of CaS can be directly related to wt % sulphur in slag, h_S can be written as $f_S \times$ (wt % sulphur in metal) and $a_{Fe} = 1$. With this simplification,

$$\text{Distribution ratio} = \frac{(\% \text{ S}) \text{ in slag}}{[\% \text{ S}] \text{ in metal}} = k' \frac{a_{CaO}}{a_{FeO}} f_S \tag{c}$$

According to Equation (c) the sulphur distribution ratio will be higher provided that: a_{CaO} is high, i.e. higher basicity; a_{FeO} is low, i.e. reducing slag; and k' is high. In Equation (c) k' increases with temperature.

Part 2

In the presence of silicon, the reaction can be written as

$$2\langle CaO \rangle + \tfrac{1}{2}[Si] + [S] = \tfrac{1}{2}\langle 2CaO \cdot SiO_2 \rangle + \langle CaS \rangle$$

for which the free energy equation is

$$\Delta G^\circ = -246\,225 + 78.0T \text{ J/mol}$$

It can be assumed, for the sake of simplicity, that the activity of lime (a_{CaO}), the activity of dicalcium silicate ($a_{2CaO \cdot SiO_2}$) and the activity of calcium sulphide (a_{CaS}), all being in the pure solid state, are unity. The equilibrium constant can then be written as

$$K = \frac{1}{[h_{Si}]^{1/2}[h_S]} = \frac{1}{[f_{Si} \text{ wt % Si}]^{1/2}[f_S \text{ wt % S}]}$$

Since $\Delta G^\circ = -RT \ln K$, hence, at 1583 K, $\ln K = 9.32$.

The Henrian activity coefficient of silicon can be obtained from

$$\lg f_{Si} = \sum e_{Si}^j (\text{wt \%})_j$$
$$= e_{Si}^S [\text{wt \% S}] + e_{Si}^C [\text{wt \% C}] + e_{Si}^{Mn} [\text{wt \% Mn}] + e_{Si}^P [\text{wt \% P}]$$
$$+ e_{Si}^{Si} [\text{wt \% Si}]$$

On substituting the interaction parameter values from Table 2.2 and the composition:

$$\lg f_{Si} = 0.056[\text{wt \% S}] + 0.775$$

Similarly the activity coefficient of sulphur (f_S) can be obtained from

$$\lg f_S = e_S^S\,[\text{wt \% S}] + e_S^C\,[\text{wt \% C}] + e_S^{Mn}\,[\text{wt \% Mn}] + e_S^P\,[\text{wt \% P}]$$
$$+ e_S^{Si}\,[\text{wt \% Si}]$$

On substituting the interaction parameter values and the composition:

$$\lg f_S = -0.028[\text{wt \% S}] + 0.527$$

On substituting the values of $\lg f_{Si}$ and $\lg f_S$ into the expression for the equilibrium constant K, we can calculate the equilibrium sulphur content of metal for a given silicon content of metal as 0.000 01 wt % or 0.1 ppm. Such low sulphur contents are not obtained in industrial practice due to the fact that the equilibrium is not achieved because of the very slow reaction kinetics at low sulphur contents.

2.4.2   *Ionic theory of slags*

According to the ionic theory of slags, oxides like MgO, CaO and FeO completely dissociate (ideal ionic behaviour) and supply oxygen ions, namely

$$MgO = Mg^{2+} + O^{2-} \tag{2.87}$$

$$CaO = Ca^{2+} + O^{2-} \tag{2.88}$$

$$FeO = Fe^{2+} + O^{2-} \tag{2.89}$$

The oxygen ions, in turn, break up the polymeric (or network) structure of SiO_2, for example

$$-O-\overset{\overset{\displaystyle O}{|}}{\underset{\underset{\displaystyle O}{|}}{Si}}-O-\overset{\overset{\displaystyle O}{|}}{\underset{\underset{\displaystyle O}{|}}{Si}}-O- + O^{2-} \rightarrow 2(SiO_4^{4-})$$

or a single molecule of silica is transformed to SiO_4^{4-} ions by two oxygen ions:

$$SiO_2 + 2O^{2-} = SiO_4^{4-} \tag{2.90}$$

Similar to SiO_4^{4-} ions, it is postulated that PO_4^{3-} and AlO_3^{3-} also form in a basic slag:

$$P_2O_5 + 3O^{2-} = 2PO_4^{3-} \tag{2.91}$$

$$Al_2O_3 + 3O^{2-} = 2AlO_3^{3-} \tag{2.92}$$

Flood and Grjotheim [4] proposed that electrically equivalent ion fractions (N') should be used instead of simple ion fractions in order to give due weight to the polarizing effect of the electrical charge on each ion present in the slag. For example, a slag consisting of w moles of SiO_2, t moles of P_2O_5, y moles of Al_2O_3 and z moles of basic oxide RO (where RO stands for MgO, CaO, etc.) will dissociate into ions according to the scheme

$$wSiO_2 + tP_2O_5 + yAl_2O_3 + zRO = wSiO_4^{4-} + 2tPO_4^{3-} + 2yAlO_3^{3-}$$
$$+ (z - 2w - 3t - 3y)O^{2-} + zR^{2+} \quad (2.93)$$

The electrically equivalent ion fraction of SiO_4^{4-} is then given by

$$N'_{SiO_4^{4-}} = \frac{4wSiO_4^{4-}}{4wSiO_4^{4-} + 6tPO_4^{3-} + 6yAlO_3^{3-} + 2(z - 2w - 3t - 3y)O^{2-}} \quad (2.94)$$

The multiplication by 4 in the numerator refers to the negative charge of 4 on silica ions (SiO_4^{4-}). The same procedure is adopted to calculate the electrically equivalent ion fractions of PO_4^{3-}, AlO_3^{3-} and O^{2-}.

If ideal ionic behaviour is assumed then the equilibrium quotient for the general reaction

$$[M] + (RO) = [O] + (RM) \quad (2.95)$$

can be written as

$$K'_R = \frac{N_{R^{2+}} N_{M^{2-}} X_O}{N_{R^{2+}} N_{O^{2-}} X_M} \quad (2.96)$$

where the ionic fractions (N) can be calculated in terms of the moles (n) of the individual cations or anions:

$$N_{R^{2+}} = \frac{n_{R^{2+}}}{\sum n_{cations}} \quad (2.97)$$

$$N_{O^{2-}} = \frac{n_{O^{2-}}}{\sum n_{anions}} \quad (2.98)$$

$$N_{M^{2-}} = \frac{n_{M^{2-}}}{\sum n_{anions}} \quad (2.99)$$

It is assumed that the elements in steel obey Henry's law and the ions in slag behave ideally.

Application of the ionic theory of slag to sulphur transfer

Consider the equilibrium between liquid iron containing dissolved sulphur and oxygen and a complex slag consisting of FeO, Na_2O, CaO, MnO, MgO, FeS, Na_2S, CaS, MnS and MgS:

$$[S] + (Fe, Na, Ca, Mn, Mg)O \rightarrow (Fe, Na, Ca, Mn, Mg)S + [O] \quad (2.100)$$

The overall reaction can be written as

$$[S] + (O^{2-}) = [O] + (S^{2-}) \tag{2.101}$$

for which the equilibrium quotient (K^*) and free energy ($\Delta G^{o, \text{overall}}$) are given by

$$K^* = \frac{N_{S^{2-}} f_O [O]}{N_{O^{2-}} f_S [S]} \tag{2.102}$$

and

$$\Delta G^{o, \text{overall}} = -RT \ln K^* \tag{2.103}$$

Let K'_{Ca} be the equilibrium constant for the reaction

$$(CaO) + [S] = (CaS) + [O] \tag{2.104}$$

and K'_{Mg} be the equilibrium constant for the reaction

$$(MgO) + [S] = (MgS) + [O] \tag{2.105}$$

K'_{Na}, K'_{Mn} and K'_{Fe} can be defined in a similar way. Since

$$\Delta G^o_{Ca} = -RT \ln K'_{Ca} \tag{2.106}$$

and

$$\Delta G^o_{Mg} = -RT \ln K'_{Mg} \tag{2.107}$$

then

$$\lg K^* = N'_{Ca} \lg K'_{Ca} + N'_{Mg} \lg K'_{Mg}$$
$$+ N'_{Fe} \lg K'_{Fe} + N'_{Na} \lg K'_{Na} + N'_{Mn} \lg K'_{Mn} \tag{2.108}$$

Application of the ionic theory of slag to phosphorus transfer

The overall dephosphorization equilibrium can be written as

$$2[P] + 5[O] + 3(O^{2-}) = 2(PO_4^{3-}) \tag{2.109}$$

for which the equilibrium quotient K^* is given by

$$K^*_{PO_4^{3-}} = \frac{(N_{PO_4^{3-}})^2}{[P]^2[O]^5(N_O^{2-})^3} \tag{2.110}$$

where N represents the anion fractions.

Phosphorus distribution data are available in the literature for a variety of experimental conditions. Equation (2.110), however, is applicable only to single-phase slags and for ideal mixing of the anions and cations. For non-ideal mixing of the ions

$$\lg K^*_{PO_4^{3-}} = [\sum N'_{Mi} \lg K_{Mi}] + f(\gamma) \tag{2.111}$$

where, as before, N'_{Mi} are the electrically equivalent ion fractions and K'_{Mi} are the true equilibrium constants for reactions of the type

$$2[P] + 5[O] + 3(MiO) = (Mi_3P_2O_8) \tag{2.112}$$

The function $f(\gamma)$ represents the sum of the terms accounting for deviation from ideal behaviour. An empirical expression for $f(\gamma)$ has been proposed [5] as follows:

$$f(\gamma) = -8.35 + 1.41S + (13\,200/T) + [40.55 - 6.95S - (64\,500/T)]N_{O^{2-}}$$
$$+ [-13.10 + 5.91S + (22\,100/T)](N_{O^{2-}})^2 \tag{2.113}$$

where S accounts for the influence of SiO_2 on the activity of FeO and CaO. The value of S is given by the ratio of ionic fractions defined as

$$S = \frac{N_{SiO_4^{4-}}}{N_{SiO_4^{4-}} + N_{PO_4^{3-}}} \tag{2.114}$$

Another relation similar to Equation (2.113) has been proposed [6] on the basis of regression analysis of industrial and experimental data:

$$f(\gamma) = -0.177\,96N_{Ca^{2+}} - 1.3787N_{Fe^{2+}} - 6.6318N_{PO_4^{3-}}$$
$$- (16\,519/T) + 2.0874V - 0.043\,36N_{Ca^{2+}} + V + (0.038\,261/T)N_{Fe^{2+}}$$
$$+ 10.905 \tag{2.115}$$

where V is the wt % ratio defined as

$$V = \frac{(wt\,\%\,P_2O_5)}{(wt\,\%\,P_2O_5) + (wt\,\%\,SiO_2)} \tag{2.116}$$

Example 2.5 **Phosphorus distribution between slag and metal**

1. What are the conditions for obtaining a high phosphorus distribution between slag and metal?
2. Calculate the equilibrium phosphorus content of 300 tons of steel held at equilibrium with 20 tons of slag containing 55 wt % CaO, 15 wt % FeO, 25 wt % SiO_2 and 5 wt % P_2O_5, at 1923 K.

    In order to obtain a better phosphorus distribution, which one of the following options is best:

    Option (a): simply lower the temperature to 1913 K?
    Option (b): increase the slag temperature to 1933 K by blowing oxygen and thereby oxidize iron such that the FeO content of the slag increases to 20 wt %?

Solution

Phosphorus distribution has been studied on both a laboratory and a plant scale [5–13]. Healy [11] represented the apparent equilibrium constant as

$$k = \frac{(wt \% P)}{[wt \% P] \times (wt \% Fe \text{ total})^{2.5}} \tag{A}$$

and

$$\lg \frac{(wt \% P)}{[wt \% P]} = (22\,350/T) - 16.0 + 0.08(wt \% CaO) + 2.5\lg(wt \% Fe_t) \tag{B}$$

The favourable effects of CaO and FeO and the detrimental effect of temperature on phosphorus distribution are evident from the above equations. The effects of MgO, CaF_2, etc., are not incorporated in Equation (B). Since most of the empirical expressions employed in steel plants are modified forms of Healy's equation, the same would be employed in the following calculations.

It is necessary to set up mass balance equations in conjunction with Equation (B) because the weight and composition of both metal and slag will change as a result of phosphorus and iron transfer. Let the final weight of slag be WSL and the final weight of metal be WM. Let the equilibrium phosphorus content of metal and slag be [wt % P] and (wt % P), respectively.

Phosphorus balance equation

$$WM \frac{[wt \% P]}{100} + WSL \frac{(wt \% P)}{100} = WP \tag{C}$$

Iron balance equation

The slag contains 20 wt % FeO. The percentage of iron in the slag (wt % Fe_t) is related to wt % FeO in the slag by

$$(wt \% Fe_t) = \frac{56}{72}(wt \% FeO)$$

Thus, for a slag containing 20 wt % FeO,

$$(wt \% Fe_t) = \frac{56}{72} \times 20 = 15.56 \text{ wt } \%$$

The iron balance equation can be written as

$$WM \frac{(100 - [wt \% P])}{100} + WSL \frac{(wt \% Fe_t)}{100} = WFe \tag{D}$$

where WFe is the total weight of iron in slag and metal.

Slag balance

Let the final weights of CaO, SiO_2, FeO and P_2O_5 be $WCaO$, $WSiO_2$, $WFeO$ and WP_2O_5, respectively. Then

$$WCaO + WSiO_2 + WFeO + WP_2O_5 = WSL \tag{E}$$

The weight of lime ($WCaO$) and weight of silica ($WSiO_2$) are known and remain unchanged:

$$WCaO = 20 \times 10^3 \times \frac{55}{100} \text{ kg}$$

$$WSiO_2 = 20 \times 10^3 \times \frac{25}{100} \text{ kg}$$

$WFeO$ can be expressed in terms of wt % FeO and WSL:

$$WFeO = (\text{wt \% FeO}) \frac{WSL}{100} \text{ kg}$$

Similarly, WP_2O_5 can be expressed in terms of (wt % P) and WSL.

Equation (B) can be rewritten in terms of $WCaO$:

$$\lg \frac{(\text{wt \% P})}{[\text{wt \% P}]} = (22\,350/T) - 16.0 + 0.08\left(\frac{WCaO}{WSL} \times 100\right)$$
$$+ 2.5 \lg(\text{wt \% Fe}_t) \tag{F}$$

The equations (C), (D), (E) and (F) constitute four non-linear equations in four unknowns, namely WSL, WM, (wt % P) and [wt % P]. The solution can be obtained on a computer to give the following results:

Option (a): Lowering the temperature from 1923 to 1913 K will result in a phosphorus distribution of

$$\lg \frac{(\text{wt \% P})}{[\text{wt \% P}]} = 2.857\,28$$

Option (b): Increasing the FeO content of slag at the expense of temperature will result in a phosphorus distribution of

$$\lg \frac{(\text{wt \% P})}{[\text{wt \% P}]} = 2.8068$$

Thus, from thermodynamic equilibrium considerations, option (a) is slightly better than option (b). However, in practice, to save processing time, option (b) is preferred. It may take a very long time for the equilibrium to establish in either case but in option (b) we can take the liberty of overoxidizing the bath to, say, 21 wt % FeO and to achieve the desired phosphorus content rather quickly. The FeO content of slag in modern combined converters lies in the range of 13–17 wt %.

2.4.3   Thermodynamic models of slags

The earliest thermodynamic models of slags were the structureless models [14–17] in which the principal reaction was represented by Equation (2.90). The oxygen in SiO_4^{4-} is single bonded (O^-). The oxygen in SiO_2 is double bonded but has no electric charge (O^0). The oxygen supplied by basic oxides like CaO has a negative charge of two (O^{2-}) and the overall reaction is written as

$$2(O^{2-}) + 2(O^0) = 4(O^-) \tag{2.117}$$

If the equilibrium laws were considered to be applicable to the above reaction

$$K' = \frac{(a_{O^-})^4}{(a_{O^{2-}})^2 (a_{O^0})^2} \tag{2.118}$$

Most of the later models made use of Equation (2.118) and can be divided into two categories, namely the models based on classical thermodynamics [18–29] and those based on statistical thermodynamics [30]. A limitation of the models based on classical thermodynamics is that they cannot be extended to ternary systems and therefore the present discussion will be limited to the latter category of models which have been applied very successfully to ternary and multicomponent slags in steelmaking [30–33].

According to statistical thermodynamics, the free energy (G) is related to the configurational partition function Ω by

$$G = -RT \ln \Omega \tag{2.119}$$

Ω is given by the total number of distinguishable configurations, $g(E_i)$, over all the energy states E_i:

$$\Omega = \sum_i g(E_i) \exp(-E_i/RT) \tag{2.120}$$

In order to find the value of Ω for slags containing SiO_2 along with two basic oxides M_IO and $M_{II}O$, the melt is divded into cells of type M_IOM_I, $M_{II}OM_{II}$, Si–O–Si, M_I–O–Si, M_{II}–O–Si and M_I–O–M_{II}. The first three cells are symmetric whereas the last three are formed according to the reactions

$$M_I\text{–O–}M_I + M_{II}\text{–O–}M_{II} = 2M_IOM_{II} \tag{2.121}$$

$$M_I\text{–O–}M_I + \text{Si–O–Si} = 2M_IOSi \tag{2.122}$$

$$M_{II}\text{–O–}M_{II} + \text{Si–O–Si} = 2M_{II}OSi \tag{2.123}$$

Yokowa and Niwa [34] assumed that the symmetric and asymmetric cells form an ideal solution. Kapoor *et al.* [35–37] derived the following generalized expressions for the activities of components by taking into account the non-ideal behaviour of cells of various types:

$$a_{M_I O} = \frac{N_I - R_{I,S} - R_{I,II}}{N_I(1 - N_S)^{1/2}} \exp\left(\frac{2[\varepsilon_{IS,SS} - \varepsilon(1 - N_S)](N_S - R_{I,S} - R_{II,S})}{RT}\right) \quad (2.124)$$

$$a_{M_{II} O} = \frac{N_{II} - R_{II,S} - R_{I,II}}{N_{II}(1 - N_S)^{1/2}} \exp\left(\frac{2[\varepsilon_{IIS,SS} - \varepsilon(1 - N_S)](N_S - R_{I,II,S})}{RT}\right) \quad (2.125)$$

$$a_{SiO_2} = \frac{(N_S - R_{I,S} - R_{II,S})^2}{N_S^3} \exp\left(\frac{4\varepsilon(1 - N_S)(1 - N_S + R_{I,S} + R_{II,S})}{RT}\right) \quad (2.126)$$

where $R_{I,S}$, $R_{II,S}$ and $R_{I,II}$ are fractions of cells of the type $M_I O Si$, $M_{II} O Si$ and $M_I O M_{II}$; N_I, N_{II} and N_S are fractions of oxygen ions supplied by the oxides $M_I O$, $M_{II} O$ and SiO_2; and $\varepsilon_{IS,SS}$ and $\varepsilon_{IIS,SS}$ are the interaction energies between bonds of the type $M_I O Si$ and Si–O–Si and $M_{II} O Si$ and Si–O–Si. The fractions of various asymmetric bonds are calculated by solving the following simultaneous set of equations:

$$(N_I - R_{I,S} - R_{I,II})(N_{II} - R_{II,S} - R_{I,II}) = R_{I,II}^2 \exp\left(-\frac{2W_{I,II}}{RT}\right) \quad (2.127)$$

$$(N_I - R_{I,S} - R_{I,II})(N_S - R_{I,S} - R_{II,S}) = R_{I,S}^2 \exp\left(\frac{2W_{I,S}}{RT}\right)\exp\left(-\frac{2\varepsilon(1 - N_S)}{RT}\right)$$

$$(2.128)$$

$$(N_{II} - R_{II,S} - R_{I,II})(N_S - R_{I,S} - R_{II,S}) = R_{II,S}^2 \exp\left(\frac{2W_{II,S}}{RT}\right)\exp\left(-\frac{2\varepsilon(1 - N_S)}{RT}\right)$$

$$(2.129)$$

where $W_{I,II}$, $W_{I,S}$ and $W_{II,S}$ are the energies of the formation of cells of the type $M_I O M_{II}$, $M_I O Si$ and $M_{II} O Si$, respectively. The average cell interaction energy, ε, is given by

$$\varepsilon = \frac{N_I \varepsilon_{IS,SS} + N_{II} \varepsilon_{IIS,SS}}{N_I + N_{II}} \quad (2.130)$$

The calculation of the cell interaction energy (ε) in a multicomponent system is simplified by making the following assumptions such that ultimately only binary systems need to be considered:

1. The interaction between two cells of the same type is zero.

2. It is assumed that

$$\varepsilon_{ji-ii} = 2\varepsilon_{ij-ii}$$

$$\varepsilon_{ij-kk} = \varepsilon_{ik-kk} + \varepsilon_{jk-kk}$$

i.e. only the interaction parameters ε_{ij-ii} are considered which will, henceforth, be denoted as ε_{ij}.

3. For systems of strong interaction (e.g. in $CaO-SiO_2$) the interaction energy is assumed to vary linearly with composition:

$$\varepsilon_{ij} = (\varepsilon_{ij})_1 + (\varepsilon_{ij})_2 X_i$$

where X_i is the mole fraction of i.

4. For calculating the total interaction energy it is assumed that the cells are randomly distributed; this only implies that the formation energy of the cells is much more than their interaction energy.

5. The interaction energy is independent of temperature.

With the above assumptions, the values of interaction parameters have been calculated for each binary system by using the available data on partial or integral molar heats of mixing. In the absence of experimental data the parameters are calculated from the respective phase diagrams. The following notation holds for multicomponent systems [38]:

$$\left.\begin{array}{l} u_i \ (i = 1, \ldots, m) \\ v_i \ (i = 1, \ldots, m) \end{array}\right\} \text{stoichiometric indices of the oxide } (M_i)_{u_i} O_{v_i}$$

$X_i \ (i = 1, \ldots, m)$, mole fractions of oxides

$R_{ij} \ (i = 1, \ldots, m; j = 1, \ldots, m)$, fraction of cells M_i-O-M_j

$\omega_{ij} \ (i = 1, \ldots, m-1; j = i+1, \ldots, m)$, formation energy of cells
$$[\omega_{ij} = (\omega_{ij})_1 + (\omega_{ij})_2 X_i]$$

$\varepsilon_{ij} \ (i = 1, \ldots, m-1; j = i+1, \ldots, m)$, interaction energy of cells
$$[\varepsilon_{ij} = (\varepsilon_{ij})_1 + (\varepsilon_{ij})_2 X_i]$$

$$D_i = \sum_{k=i}^{m} v_k X_k$$

$$R_{ij}^* = \frac{v_i v_j X_i X_j}{D_i}, \text{ fraction of cells } M_i-O-M_j \text{ for a completely random distribution of cations}$$

$$Q_i = \prod_{k=i+1}^{m} \exp\left(\frac{v_k X_k}{D_i} \cdot \frac{\varepsilon_{ik}}{RT}\right)$$

$$P_{ij} = \begin{cases} Q_i Q_j \exp(-\omega_{ij}/RT) & \text{if } j \neq i \\ 1 & \text{if } j = i \end{cases}$$

The value of ΔG^M (integral molar free energy of mixing), assuming the constituents to be in the same state as the solution, can be calculated from

$$\Delta G^M = 2 \sum_{i=1}^{m-1} \sum_{j=i+1}^{m} \left(R_{ij}\omega_{ij} + R_{ii}\frac{v_j X_j}{D_1}\varepsilon_{ij} \right)$$

$$- RT \sum_{i=1}^{m-1} \frac{u_i}{v_i}\left[D_i \ln\left(\frac{D_i}{v_i X_i}\right) - D_{i+1} \ln\left(\frac{D_{i+1}}{v_i X_i}\right) \right]$$

$$- RT \sum_{i=1}^{m} \sum_{j=1}^{m} (R_{ij}^* \ln R_{ij}^* - R_{ij} \ln R_{ij}) \tag{2.131}$$

The fraction of cells R_{ij} are subject to $m(m+1)/2$ constraints as follows.
    For i varying from 1 to m:

(a) $\displaystyle\sum_{j=1}^{m-1} R_{ij} = v_i X_i$ $\hspace{6cm}$ (2.132)

(b) $R_{ii}R_{jj} - R_{ij}^2/P_{ij}^2 = 0$ $\quad (j = i+1, \ldots, m)$ $\hspace{3cm}$ (2.133)

The relationships grouped in Equation (2.132) are linear while those in Equation (2.133), arising from the maximization of the partition function with respect to the number of cells, are quadratic. Depending upon the number of constituents, the number of linear and non-linear equations to be solved are as shown below.

Number of constituents	Number of relationships in Equations (2.132) and (2.133)		
	Linear	Quadratic	Total
2	2	1	3
3	3	3	6
4	4	6	10
5	5	10	15
6	6	15	21

The convergence of the solution becomes progressively difficult as the number of equations increase. However, with the substitution $y_i = \sqrt{(R_{ii})}$ we obtain only m equations in terms of y_i:

$$y_i = \sum_{j=1}^{m} P_{ij} Y_j = v_i X_i \quad (i = 1, \ldots, m) \tag{2.134}$$

After obtaining the solution of m equations in m unknowns, the R_{ij} are calculated from

$$R_{ij} = P_{ij} \ldots Y_i \ldots Y_j \quad (i = 1, \ldots, m; j = 1, \ldots, m) \tag{2.135}$$

These values are then substituted into Equation (2.131) to obtain ΔG^M. The ΔS^M and ΔH^M are calculated from the equations

$$\Delta S^M = \frac{\Delta G^M(T - \Delta T) - \Delta G^M(T + \Delta T)}{2\Delta T} \tag{2.136}$$

$$\Delta H^M = \Delta G^M + \Delta T S^M \tag{2.137}$$

The activity is calculated from

$$RT \ln a_i = \Delta G^M + \sum_{j=1}^{m-1} (\delta_{ij} - X_j) \frac{\partial \Delta G^M}{\partial X_j} \quad (i = 1, \ldots, m) \tag{2.138}$$

where δ_{ij} represent the Kronecker delta symbol.

At a given temperature T, the activities in two different states are related approximately by

$$\frac{(a_{MO})_s}{(a_{MO})_l} = \exp\left(\frac{\Delta H^{fuse} - T\Delta S^{fuse}}{RT}\right) \tag{2.139}$$

In Table 2.3, the values of parameters $(\omega_{ij})_1$ and $(\omega_{ij})_2$ as well as $(\varepsilon_{ij})_1$ and $(\varepsilon_{ij})_2$ are summarized.

A computer program of the Kapoor–Frohberg model is used extensively [35–37] to calculate activities as well as liquidus temperatures for the known

Table 2.3   Energy parameters used in the model.

System	Cation i	j	Cells formation (Joules) $(\omega_{ij})_1$	$(\omega_{ij})_2$	Cells interaction (Joules) $(\varepsilon_{ij})_1$	$(\varepsilon_{ij})_2$
SiO_2–Al_2O_3	Si	Al	8368	0	−12 552	0
SiO_2–Fe_2O_3	Si	Fe^{3+}	4184	0	6694	0
SiO_2–CaO	Si	Ca	−52 300	0	−18 828	31 380
SiO_2–FeO	Si	Fe^{2+}	−6276	0	8786.4	4184
SiO_2–MgO	Si	Mg	−33 472	0	5020.8	12 552
SiO_2–MnO	Si	Mn	−18 828	0	−4184	21 756.8
Al_2O_3–CaO	Al	Ca	−35 564	12 552	−23 012	−20 920
Al_2O_3–FeO	Al	Fe^{2+}	−1673.6	0	−9623.2	0
Al_2O_3–MgO	Al	Mg	−14 644	0	−29 288	0
Al_2O_3–MnO	Al	Mn	−5020.8	0	−13 388.8	0
Fe_2O_3–CaO	Fe^{3+}	Ca	−30 961.6	0	−8368	0
Fe_2O_3–FeO	Fe^{3+}	Fe^{2+}	−2092	0	−2092	0
Fe_2O_3–MnO	Fe^{3+}	Mn	−2092	0	−2092	0
CaO–FeO	Ca	Fe^{2+}	−12 552	0	2092	0
CaO–MgO	Ca	Mg	4182	0	0	0
CaO–MnO	Ca	Mn	−2092	0	−3347.2	0
FeO–MgO	Fe^{2+}	Mg	−2092	0	0	0
FeO–MnO	Fe^{2+}	Mn	−2092	0	0	0
MgO–MnO	Mg	Mn	−2092	0	−3347.2	0

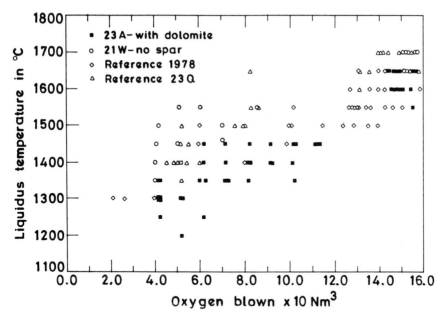

Figure 2.5 Liquidus temperatures of slags during a heat in a combined-blown converter at Hoogovens IJmuiden; heats with dolomite show lower temperature compared with heats with lime only.

compositions of steelmaking slags. As an example, in Figure 2.5, the calculated liquidus temperatures [39] are shown for dolomitic and non-dolomitic slags in 315 ton converters at Hoogovens with blowing time; the liquidus temperatures are lowered by about 100 K because of the fluxing effect of MgO.

2.4.4 Thermodynamic criteria for the refining capacity of slags

The oxygen potential corresponding to FeO is usually the highest in steelmaking slags and thus FeO controls the oxidizing power of the slag phase; if $p_{O_2}^o$ is the equilibrium partial pressure in contact with pure iron oxide

$$(p_{O_2}/p_{O_2}^o)^{1/2} = a_{FeO} \tag{2.140}$$

Since $p_{O_2}^o$ is constant, the activity of FeO is directly proportional to $(p_{O_2})^{1/2}$. The iso-activity lines of FeO (Figure 2.6) show maxima close to the line joining the FeO$_n$ apex to the dicalcium silicate (C$_2$S) composition on the CaO–SiO$_2$ line. At higher silica contents (i.e. basicity in the range of 1–2) the activity of FeO decreases due to the formation of iron silicate. With the increase of lime (basicity greater than 2) the activity of FeO again decreases because it combines with the lime to form ferrites. Since the oxidizing power of the slag for a given FeO content is at a maximum

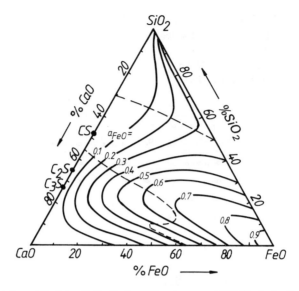

Figure 2.6   Activity of FeO in CaO–FeO–SiO$_2$ slags at 1873 K; C$_2$S, dicalcium silicate; C$_3$S, tricalcium silicate. (*Source*: Darken, L. S. and Gurry, R. W., *J. Am. Chem. Soc.*, **67** (1945) 1398–412; **68** (1946) 798–816.)

when the basicity is approximately 2, in practice slags of basicity 2–3.5 are employed for the effective elimination of phosphorus. There is an upper limit to the desirable amount of FeO in the slag at a constant basicity; at high FeO contents (say, above 20%) the FeO physically replaces CaO in the slag and adversely affects the phosphorus partition which is governed by the reaction

$$2[P] + 5(FeO) + 3(CaO) = (3CaO \cdot P_2O_5) + 5\{Fe\}$$

The FeO in the slag (oxidizing power) is required to oxidize phosphorus but CaO is required (i.e. higher basicity) to arrest the P$_2$O$_5$ in the slag as $3CaO \cdot P_2O_5$. Both basicity and oxidizing power, in spite of their practical utility, are qualitative terms; an optimum combination of basicity and oxidizing power will vary with slag composition.

A quantitative thermodynamic description of the refining capacity of slags can be provided in terms of sulphide capacity, phosphate capacity and optical basicity.

Sulphide capacity

Consider the equilibrium constant for the reaction

$$\tfrac{1}{2}(S_2)_g + (O^{2-}) = (S^{2-}) + \tfrac{1}{2}(O_2)_g$$

The equilibrium constant can be written as

$$K_{S^{2-}} = \left(\frac{a_{S^{2-}}}{a_{O^{2-}}}\right)\left(\frac{p_{O_2}}{p_{S_2}}\right)^{1/2} = \frac{X_{S^{2-}}\gamma_{S^{2-}}}{a_{O^{2-}}}\left(\frac{p_{O_2}}{p_{S_2}}\right)^{1/2} \tag{2.141}$$

Since the concentration of oxygen ions in basic slags is high and the sulphur content is small, both $a_{O^{2-}}$ and $\gamma_{S^{2-}}$ can be treated as constants for a given oxide content of the slag. Thus, on converting $X_{S^{2-}}$ to (wt % S), one can write

$$C_S = (\text{wt \% S})\left(\frac{p_{O_2}}{p_{S_2}}\right)^{1/2} \tag{2.142}$$

where

$$C_S = \frac{K_{S^{2-}}a_{O^{2-}}}{\gamma_{S^{2-}}} \tag{2.143}$$

C_S is known as the sulphide capacity and is obtained from gas–slag equilibria. For metal–slag equilibria, the apparent equilibrium constant can be written as

$$[S] + (O^{2-}) = (S^{2-}) + [O]$$

i.e.

$$k_s = (\text{wt \% S})(h_O/h_S) \tag{2.144}$$

From the free energy of a solution of gaseous oxygen and sulphur in liquid iron

$$[S] + \tfrac{1}{2}(O_2)_g = \tfrac{1}{2}(S_2)_g + [O] \tag{2.145}$$

$$K_{OS} = \frac{[h_O]}{[h_S]}\left(\frac{p_{S_2}}{p_{O_2}}\right)^{1/2} \tag{2.146}$$

$$\lg K_{OS} = -(935/T) + 1.375 \tag{2.147}$$

At low concentrations of sulphur in iron, $[h_S] = [\text{wt \% S}]$ and k_s can be calculated in terms of C_S:

$$k_s = C_S K_{OS} \tag{2.148}$$

It is thus possible to calculate k_s from the experimentally measured values of sulphide capacity, C_S.

Data for the sulphide capacity are available for a wide range of compositions of blast furnace type of slags [40]. In general, C_S increases with temperature by a factor of 1.25 for each 50 K rise in temperature [41]. Similar slag compositions are also employed in ladle metallurgical treatments. The polymer theory of slags has been applied [42] to calculate the sulphide capacity in binary $CaO-SiO_2$ and $FeO-SiO_2$ slags as well as in ternary $CaO-FeO-SiO_2$ slags (Figure 2.7). The calculated and measured values at 1773 K for constant mole fractions of SiO_2 agree well [42].

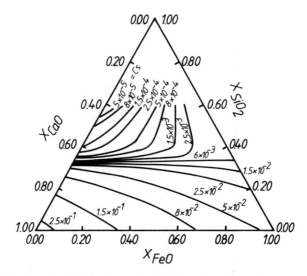

Figure 2.7   Calculated sulphide capacity (C_S) in the ternary CaO–FeO–SiO$_2$ system. (*Source*: Chen, B., Reddy, R. G. and Blander, M. (1989) *Proc. 3rd Int. Conf. on Molten Slags and Fluxes, Glasgow, 27–29 June 1988*, The Institute of Metals: London, pp. 270–2.)

Phosphate capacity

Wagner [43] introduced the concept of phosphate capacity, similar to sulphide capacity. Consider the gas–slag reaction

$$(P_2)_g + \tfrac{5}{2}(O_2)_g + 3(O^{2-}) = 2(PO_4^{3-}) \tag{2.149}$$

for which the equilibrium constant can be written as

$$K_{PO_4^{3-}} = (a_{PO_4^{3-}})^2 [(p_{P_2})(p_{O_2})^{2.5}(a_{O^{2-}})^3]^{-1} \tag{2.150}$$

For a small concentration of PO$_4^{3-}$ ions and a relatively large concentration of O^{2-} ions in steelmaking slags

$$C_{PO_4^{3-}} = \frac{(\text{wt \% PO}_4^{3-})}{(p_{P_2})^{1/2}(p_{O_2})^{1.25}} \tag{2.151}$$

In terms of the equilibrium constant

$$C_{PO_4^{3-}} = \left(\frac{a_{O^{2-}}}{f_{PO_4^{3-}}}\right)\sqrt{(K_{PO_4^{3-}})} \tag{2.152}$$

Thus, the phosphate capacity ($C_{PO_4^{3-}}$) is a measure of the dephosphorizing capability of a slag under a given oxygen potential and is related to the oxygen ion activity in the slag.

Some other definitions of phosphate capacity are

$$C_P = \frac{(\text{wt \% P}_2\text{O}_5)}{[a_P][a_O]^{2.5}} \tag{2.153}$$

$$C_{P_M} = \frac{(\text{wt \% P})}{[\text{wt \% P}][\text{wt \% O}]^{2.5}} \tag{2.154}$$

and

$$C_{PO_4^{3-}} = \frac{X_{PO_4^{3-}}}{[\text{wt \% P}](X_{FeO})^{2.5}} \tag{2.155}$$

but they can all be shown to be related to Equation (2.151).

For metal–slag equilibria, it is customary to define phosphorus distribution as $(\text{wt \% P})/[\text{wt \% P}]$. The solubility of phosphorus in liquid iron is

$$\tfrac{1}{2}(P_2)_g = [P]_{1\,\text{wt \%}} \quad \Delta G° = -122\,173 - 19.25T \quad \text{J/mol} \tag{2.156}$$

Similarly, for oxygen

$$\{Fe\} + \tfrac{1}{2}(O_2)_g = (FeO) \quad \Delta G° = -264.64 + 0.065T \quad \text{J/mol} \tag{2.157}$$

From Equations (2.151) and (2.155)–(2.157)

$$\lg((P)/[P]) = \lg C_{PO_4^{3-}} - (41\,113/T) + 7.06 + 2.5 \lg a_{FeO} + \lg f_P \tag{2.158}$$

The phosphate capacity depends strongly on temperature and the slag composition. The data for the phosphate capacity of Fe_2O_3–FeO–CaO–SiO_2–P_2O_5 slags are summarized in Figure 2.8 as a function of temperature and composition.

Optical basicity

The concept of optical basicity was initially developed by Duffy and Ingram [44] for glasses. It is based on the following:

1. The spectroscopic shift observed during the 6s to 6p transition of those metal ions (i.e. Ti^{2+}, Pb^{2+}, Bi^{3+}) which are present in oxidation states two units less than the number of the group to which they belong.
2. The basicity of the oxides present in glasses.

Since the spectroscopic shift represents the electron donor power of the oxides in a glass, one glass may be compared with another simply by comparing the basicity ratios expressed as electron donor power of the oxides in the glass to the electron donor power of the free oxide ion. The ratio of donor powers derived from spectroscopic measurements was termed optical basicity, Λ, by Duffy and Ingram. It was later found [45] that the optical basicity of an oxide AO, written as Λ_A, was

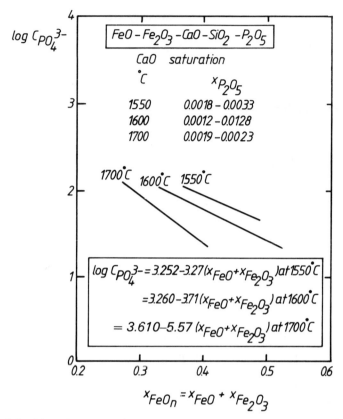

Figure 2.8 Phosphate capacity in Fe_2O_3–FeO–CaO–SiO_2–P_2O_5 slags. (*Source*: Wrampelmeyer, J.-C., Dimitrov, S. and Janke, D. (1989) *Steel Res.*, **60**, 539–49.)

related to the Pauling electronegativity, ξ, of the cation:

$$\Lambda_A = \frac{0.74}{(\xi - 0.26)} \tag{2.159}$$

For a mixture of oxides one can calculate an average value of optical basicity:

$$\Lambda = X'_A \Lambda_A + X'_B \Lambda_B + \cdots \tag{2.160a}$$

where

$$X'_A = \frac{(\text{mole fraction of component A})(\text{no. of oxygen ions in the molecule})}{\sum (\text{mole fraction of components})(\text{no. of oxygen ions in the molecule})}$$

The sulphide capacity of slags has been related to optical basicity [46] as follows:

$$\lg C_S = \frac{22\,690 - 54\,640\Lambda}{T} + 43.6\Lambda - 25.2 \tag{2.160b}$$

On substituting C_S in terms of the sulphur distribution ratio $(S)/[h_S]$:

$$\lg\left(\frac{(S)}{h_S}\right) = \left(\frac{21\,920 - 54\,640\Lambda}{T}\right) + 43.6\Lambda - 23.9 - \lg[a_O] \qquad (2.161)$$

where $h_S = f_S[S]$. The value of f_S, depending upon the carbon and silicon contents of molten iron produced in the blast furnace, varies in the range 5–7; in plain carbon liquid steel the value of f_S is approximately 1.

The optical basicity has been correlated [47] to phosphorus capacity (expressed as C_{P_M} in Equation (2.154)):

$$\lg C_{P_M} = 21.55\Lambda + (32\,912/T) - 27.90 \qquad (2.162)$$

The oxygen content of metal in terms of optical basicity is expressed as [47]:

$$\lg[\text{wt \% O}] = -1.907\Lambda - (6005/T) + 3.570 \qquad (2.163)$$

The optical basicity is emerging as a new tool to predict both the sulphide and the phosphate capacity of complex slags.

2.5 Multicomponent–multiphase equilibrium calculations involving slag, gas and liquid iron alloys

The reactions in steelmaking can be divided into two major groups:

1. Oxide-forming reactions, namely

$$2[P] + 5[O] = (P_2O_5) \qquad (2.164)$$

$$[Fe] + [O] = (FeO) \qquad (2.165)$$

$$[Mn] + [O] = (MnO) \qquad (2.166)$$

$$[Si] + 2[O] = (SiO_2) \qquad (2.167)$$

$$[C] + [O] = (CO)_g \qquad (2.168)$$

2. Sulphide-forming reactions, namely

$$[Fe] + [S] = (FeS) \qquad (2.169)$$

$$[Mn] + [S] = (MnS) \qquad (2.170)$$

$$[Ca] + [S] = (CaS) \qquad (2.171)$$

$$[Si] + [S] = (SiS)_g \qquad (2.172)$$

$$[Si] + 2[S] = (SiS_2) \qquad (2.173)$$

If the alloying elements like chromium, nickel, titanium and aluminium are dissolved in liquid iron, the corresponding sulphide and oxide reactions for each element should be added. Besides the chemical composition of the phases, the multicomponent equilibrium calculations are also important to determine the actual equilibrium amounts of each phase.

Suppose a large quantity of carbon-saturated iron is brought into contact with a small quantity of reducible slag. The change in slag composition and weight can be substantial when the ultimate equilibrium is established, e.g. due to the reduction of SiO_2 in a CaO–SiO_2 slag by carbon in metal. Similarly, the change in metal composition may be significant if a small amount of metal is brought into contact with a large amount of slag. This occurs, for example, in the case of an oxygen steelmaking converter in which the molten-metal droplets ejected by the oxygen jet traverse through both the gas and the pool of liquid slag before they drop back into the metal. It is demonstrated in this section that problems of such a nature can be solved (a) by the insertion of the free energy, interaction parameter, activity and other thermodynamic data for relevant oxide and sulphide reactions and (b) by specifying the initial weights and compositions of the slag, metal and gas phases in a set of simultaneous equations. The final weights and composition of different phases at equilibrium can then be obtained by solving the simultaneous equations. For the sake of simplicity, the non-stoichiometry of oxides or sulphides has not been taken into account in the treatment that follows. The effects of non-stoichiometry are usually small. For example, ferrous oxide is non-stoichiometric and its composition is usually stated as Fe_xO where x ranges from 0.97 to 1.00. The Fe–O phase diagram shows that the value of x in equilibrium with iron at 1773 K is 0.97 and at 1873 K it is 0.98. The value of x is also a function of oxygen potential in the slag–metal–gas system. In a complex slag, the oxygen potential decreases with a decrease in the activity of FeO. However, since the overall variation in x is small, the equilibrium oxygen content is affected only marginally. A constant value of $x = 1$ is used. The errors arising from this assumption are much less than the uncertainties in the experimentally determined activities of FeO in complex steelmaking slags.

Initialization

Let the liquid metal initially contain MCA moles of carbon, MSU moles of sulphur, MNT moles of nitrogen, MHY moles of hydrogen, MOX moles of oxygen and Mi moles of other elements where i stands for iron, phosphorus, silicon, manganese, aluminium, etc. Let TMETAL and TSLAG be the total moles of metal and slag at equilibrium and let the corresponding final weights of metal and slag be FMETAL and FSLAG. XC, XS, XOX, XN and XH are the mole fractions of carbon, sulphur, oxygen, nitrogen and hydrogen and XMi are the mole fractions of other elements in metal at equilibrium where, as before, i stands for iron, silicon, manganese, aluminium, etc. p_{CO_2}, p_{CO}, p_{H_2}, p_{N_2} are the partial pressures of gases and XMi_αS$_\beta$ and XMi_pO$_\sigma$ represent the mole fractions of the ith sulphide and oxide, respectively.

Equilibrium constants for sulphides and oxides

Let the formation of sulphides be represented by the general equation

$$\alpha Mi + \beta S = Mi_\alpha S_\beta$$

where α and β are the stoichiometric constants. The equilibrium constant KSU_i can be written as

$$KSU_i = \frac{\gamma_{Mi_\alpha S_\beta} X Mi_\alpha S_\beta}{(f_{Mi} X Mi)^\alpha (f_S XS)^\beta} \qquad (2.174)$$

where γ stands for the activity coefficient of sulphide in the slag phase and f for activity coefficient (at infinite dilution) in the metal phase, except for iron which is the solvent. For steelmaking slags the activity coefficient $(\gamma_{Mi_\alpha S_\beta})$ can be obtained from the activity–composition diagram; for sulphides, however, the data are very limited.

Thus

$$\gamma_{Mi_\alpha S_\beta} = f \text{ (activity–composition diagram)} \qquad (2.175)$$

Similarly, for the formation of oxide,

$$\rho Mi + \sigma O = Mi_\rho O_\sigma$$

and

$$KOX_i = \frac{(\gamma_{Mi_\rho O_\sigma})(X Mi_\rho O_\sigma)}{(f_{Mi} X Mi)^\rho (f_O XOX)^\sigma} \qquad (2.176)$$

where KOX_i is the equilibrium constant for the formation of oxide i. The activity coefficient $\gamma_{Mi_\rho O_\sigma}$ can be obtained from the Kapoor–Frohberg model (discussed in section 2.4.3) by using published data. Thus

$$\gamma_{Mi_\rho O_\sigma} = f \text{ (activity–composition diagram)} \qquad (2.177)$$

and

$$\ln f_{Mi} = \sum^j \varepsilon_i^j X Mj + \sum^m \varepsilon_i^m Xm \qquad (2.178)$$

where m stands for carbon, oxygen, sulphur, hydrogen and nitrogen, and ε_i^j represents the interaction coefficient of element j on i.

The individual mass balance for an element i can be written as

$$[XMi]TMETAL + [\rho(XMi_\rho O_\sigma) + \alpha(XMi_\alpha S_\beta)]TSLAG = Mi \qquad (2.179)$$

It may be noted that, in general, for each element i we have to write six equations, namely Equations (2.174)–(2.179), provided that the element forms only one oxide and one sulphide. The number of equations will be less if a particular element does not form any oxide or sulphide. For example, in the case of phosphorus and aluminium we will need to write only four equations, namely Equations

(2.176)–(2.179). If a particular oxide or sulphide is in gaseous phase, e.g. SiS, then the corresponding equations for the equilibrium constants have to be written in terms of the partial pressures of the vapour species.

For carbon, hydrogen, nitrogen, oxygen and sulphur, which are invariably present as solutes in liquid iron, the following mass balance equations can be written:

- mass balance of sulphur

$$\sum^i \beta(XMi_\alpha S_\beta)(\text{TSLAG}) + XS(\text{TMETAL}) = \text{MSU} \tag{2.180}$$

- mass balance of oxygen

$$XOX(\text{TMETAL}) + \sum^i \sigma(XMi_\rho O_\sigma)(\text{TSLAG}) = \text{MOX} \tag{2.181}$$

- mass balance of carbon

$$XC(\text{TMETAL}) + n\text{CO} + n\text{CO}_2 = \text{MCA} \tag{2.182}$$

where $n\text{CO}$ and $n\text{CO}_2$ are moles of CO and CO_2 that are produced.

- mass balance of nitrogen and hydrogen

$$XN(\text{TMETAL}) + 2n\text{N}_2 = \text{MNT} \tag{2.183}$$

$$XH(\text{TMETAL}) + 2n\text{H}_2 = \text{MHT} \tag{2.184}$$

$n\text{CO}$, $n\text{CO}_2$, $n\text{H}_2$ and $n\text{N}_2$ are related to the partial pressures of gases in a closed system as follows:

$$p_{\text{H}_2} = \frac{n\text{H}_2}{n\text{CO} + n\text{CO}_2 + n\text{H}_2 + n\text{N}_2} \tag{2.185}$$

$$p_{\text{N}_2} = \frac{n\text{N}_2}{n\text{CO} + n\text{CO}_2 + n\text{H}_2 + n\text{N}_2} \tag{2.186}$$

$$p_{\text{CO}} = \frac{n\text{CO}}{n\text{CO} + n\text{CO}_2 + n\text{H}_2 + n\text{N}_2} \tag{2.187}$$

$$p_{\text{CO}_2} = \frac{n\text{CO}_2}{n\text{CO} + n\text{CO}_2 + n\text{H}_2 + n\text{N}_2} \tag{2.188}$$

If there are any other gaseous species present, they can also be included. The above partial pressures are related to the equilibrium constants

$$2[\text{H}] = (\text{H}_2)_{\text{g}}$$

and

$$KH = \frac{p_{\text{H}_2}}{(XHf_{\text{H}})^2} \tag{2.189}$$

where

$$\ln f_H = \sum^j \varepsilon_H^j X M_j + \sum^m \varepsilon_H^m X m \tag{2.190}$$

and m stands for hydrogen, carbon, sulphur, oxygen and nitrogen and j for silicon, manganese, aluminium and phosphorus (except iron). Similarly

$$2[N] = (N_2)_g$$

and

$$KN = \frac{p_{N_2}}{(XN f_N)^2} \tag{2.191}$$

where

$$\ln f_N = \sum^j \varepsilon_N^j X M j + \sum^m \varepsilon_N^m X m \tag{2.192}$$

and m stands for nitrogen, carbon, hydrogen, oxygen and sulphur and j for silicon, manganese, aluminium and phosphorus (except iron).

For the formation of carbon monoxide gas

$$[C] + [O] = (CO)_g$$

$$KCO = \frac{p_{CO}}{f_C X C f_O X O X} \tag{2.193}$$

and

$$\ln f_C = \sum^j \varepsilon_C^j X M j + \sum^m \varepsilon_C^n X m \tag{2.194}$$

where m and j are as before, and

$$\ln f_O = \sum^j \varepsilon_O^j X M j + \sum^m \varepsilon_O^m X m \tag{2.195}$$

Also

$$[C] + 2[O] = (CO_2)_g$$

and

$$KCO_2 = \frac{p_{CO_2}}{(f_O X O X)^2 (f_C X C)} \tag{2.196}$$

In Equation (2.174), the activity coefficient for sulphur is given by

$$\ln f_S = \sum^j \varepsilon_S^j X M j + \sum^m \varepsilon_S^m X m \tag{2.197}$$

where m and j are as before.

The overall mass balance equation can now be written for the metal and slag phases.

For the metal phase:

$$XC + XS + XH + XN + XOX + \sum_{}^{i} XMi = 1 \tag{2.198}$$

and

$$12XC + 32XS + 1XH + 14XN + 16XO + \sum_{}^{i} (ATi)XMi$$

$$= FMETAL/TMETAL \tag{2.199}$$

where ATi is the atomic weight of element i, i.e. iron, silicon, aluminium, manganese, phosphorus, etc.

For the slag phase:

$$\sum_{}^{i} XMi_{\alpha}S_{\beta} + \sum_{}^{i} XMi_{\rho}O_{\sigma} = 1 \tag{2.200}$$

and

$$\sum_{}^{i} XMi_{\alpha}S_{\beta}(MTSUi) + \sum_{}^{i} XMi_{\rho}O_{\sigma}(MTOXi) = FSLAG/TSLAG \tag{2.201}$$

where MTOXi and MTSUi are, respectively, the molecular weights of oxides and sulphides.

The set of 28 equations, namely Equations (2.174)–(2.201), as described above, constitute $6i + 22$ simultaneous, non-linear equations in $6i + 22$ unknowns. For example, if $i = 1$, i.e. the metal contains only iron, carbon, hydrogen, nitrogen, oxygen and sulphur, then one will need to solve 28 equations in 28 unknowns. If any of the five elements, carbon, hydrogen, oxygen, nitrogen, sulphur, are not present, then the number of equations is less. The number of equations is further reduced if equilibrium with sulphur is not considered or if the Henrian behaviour of solutes is assumed, in which case $f_C = f_H = f_O = f_N = f_S = 1$. The larger the number of equations, the greater will be the computation time required; there is almost an exponential rise in computer time with the number of equations to be solved. A number of fast converging numerical routines are available for obtaining a solution of such problems. The computer program of the Kapoor–Frohberg model is usually employed as a subroutine to estimate the activities of components in complex slag systems.

2.6   Thermodynamics of deoxidation of steel

The solubility of oxygen in pure molten iron decreases with temperature and can be calculated from Equation (2.61). At normal steelmaking temperatures the solubility is above 0.2 wt %. At the monotectic temperature in the Fe–O phase diagram it is

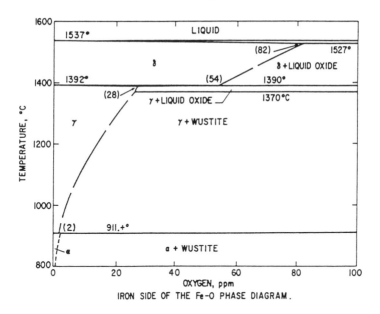

Figure 2.9 Portion of Fe–O phase diagram. (*Source*: Swisher, J. H. and Turkdogan, E. T. (1967) *Trans. Metall. Soc. AIME*, **239**, 426–31.)

0.168 wt % and is negligibly small in solid iron (Figure 2.9). Therefore, the oxygen content of steel must be reduced to a desired level while the steel is still liquid. Otherwise, during cooling and solidification, the oxygen may react with the alloying elements and form oxides. The removal of oxygen may be accomplished by adding to liquid steel those elements which have a higher affinity towards oxygen than iron. This process is known as precipitation deoxidation. The primary deoxidation products thus formed can be solid or liquid. The deoxidation products can float up if sufficient time is available and they can thus escape from the melt. The dissolved oxygen may diffuse to a suitable slag covering the steel surface. This is known as diffusion deoxidation. Deoxidation reactions may also occur during the cooling of liquid and solid steel due to segregation which leads to an increase in the concentration of the alloying elements at the boundary of the solid and liquid phases. This is called secondary deoxidation. The secondary deoxidation products usually become entrapped in solid steel and are thus retained as inclusions.

The oxides entrapped in solidified steel may be classified as indigenous or exogenous inclusions. The indigenous inclusions result from the entrapment of primary deoxidation products, whereas the exogenous inclusions are caused by the entrapment of refractory materials, including slag or scum. The exogenous inclusions have a complex chemical composition (corresponding to slag or refractory material) and are generally larger in size than those resulting from the deoxidation products. Large inclusions may also form due to the reoxidation of the deoxidized steel by air during tapping or teeming. The deoxidation products already present in liquid steel grow in size by acting as favourable sites for the reaction between the dissolved element and the oxygen from the air. The size, shape and distribution of inclusions affect the mechanical properties and the machinability of steel; see Chapter 8 for inclusion modification etc. This section is devoted to the thermodynamics of the formation of oxides in liquid steel.

The equilibrium composition of metal as well as the composition and amount of deoxidation products can be calculated by the generalized procedure described in section 2.5. An important objective in primary deoxidation is to form not only a stable but also a liquid deoxidation product which can coalesce and easily float out as well.

2.6.1   *Deoxidation by carbon*

Carbon is invariably present as an alloying element in almost all steels. Two types of reactions involved in the deoxidation by carbon are

$$[C] + [O] = (CO)_g$$

and

$$(CO_2)_g + [C] = 2(CO)_g$$

The solubility of oxygen in γ-iron is approximately 0.003 wt %. If the carbon content is less than 0.1 wt %, most of the excess oxygen combines with the carbon to form carbon monoxide during solidification. Only a part of this carbon monoxide gas escapes and the rest is entrapped in the solidified metal and compensates for the solidification shrinkage and/or causes blow holes. The evolution of the gas is small when the carbon content is greater than 0.1 wt %. At carbon contents greater than 2 wt %, the solubility of oxygen in liquid steel approaches that in the solid state. The presence or the addition of deoxidizers stronger than carbon may altogether suppress the carbon–oxygen reaction. However, if the oxygen content is still larger than that in equilibrium with carbon, then the carbon–oxygen reaction can recommence. The equilibrium C–CO–CO$_2$–O relationships in liquid steel at 1873 K, for different values of carbon monoxide pressure and the ratio of CO/CO_2, are shown in Figure 2.10 [48]. In practice a low partial pressure of carbon monoxide gas can be achieved by subjecting the liquid steel to vacuum and/or by purging the liquid steel with an inert gas like argon. Since the deoxidation product is gaseous, it escapes from the liquid steel and leaves no inclusions. In most cases, however, it becomes necessary to add deoxidizers stronger than carbon. For example, aluminium is added to modify the austenite grain size, and rare earth elements like cerium 'mischmetal' are added to modify the shape and morphology of inclusions.

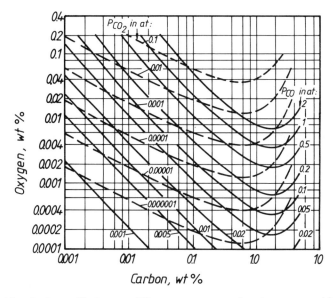

Figure 2.10   C–O equilibrium at different pressures of carbon monoxide and carbon dioxide gases. (*Source*: Steinmetz, E. and Lindenberg, H.-U. (1970) *Stahl und Eisen*, **90**, 1517–25.)

Example 2.6 **Dissolved oxygen in liquid steel in equilibrium with carbon monoxide gas**

Suppose liquid steel containing 0.5 wt % carbon at 1873 K is in equilibrium with carbon monoxide gas at 1 atm. It is required to calculate the equilibrium dissolved oxygen content, given that

$\langle C \rangle_{\text{graphite}} + \frac{1}{2}(O_2)_g = (CO)_g$		$\Delta G° = -112\,877 - 86.514T$ J/mol	
$[O]$	$= \frac{1}{2}(O_2)_g$	$\Delta G° = 117\,152 + 2.887T$ J/mol	
$[C]$	$= \langle C \rangle_{\text{graphite}}$	$\Delta G° = -22\,594 + 42.258T$ J/mol	

$$[C] \qquad + [O] \quad = (CO)_g \qquad \Delta G° = \ -18\,319 - 41.369T \text{ J/mol}$$

The equilibrium constant for the above reaction is given by

$$\ln K_{CO} = -\Delta G°/RT = 6.152$$

Or

$$K_{CO} = 469.8 = \frac{p_{CO}}{h_C h_O} = \frac{p_{CO}}{f_C[C]f_O[O]}$$

The values of the activity coefficients can be calculated as follows:

$$\lg f_C = e_C^C [wt \% C] + e_C^O [wt \% O] = 0.14[0.5] - 0.26[wt \% O]$$

$$\lg f_O = e_O^O [wt \% O] + e_O^C [wt \% C] = -0.20[wt \% O] - 0.35[0.5]$$

Now

$$\lg K_{CO} = \lg p_{CO} - \lg f_C - \lg[wt \% C] - \lg[wt \% O] - \lg f_O$$

On substituting the values of $\lg f_C$, $\lg f_O$ and $[wt \% C]$ and solving for $[wt \% O]$:

$$[wt \% O] = 0.0055\% \text{ or } 55 \text{ ppm}$$

Example 2.7 C–CO equilibrium in alloy steel

Let the composition of liquid steel (in wt %) in Example 2.6 be changed to

C	Si	Mn	S	P	Cr	O	Fe
0.5	0.2	0.2	0.03	0.02	0.2	(unknown)	rest

The new value of the activity of carbon can be calculated from

$$\lg f_C = e_C^C [wt \% C] + e_C^S [wt \% S] + e_C^{Mn} [wt \% Mn] + e_C^{Si} [wt \% Si]$$
$$+ e_C^P [wt \% P] + e_C^{Cr} [wt \% Cr] + e_C^O [wt \% O]$$

On substituting the appropriate values of the interaction parameters and following the same procedure of calculation as in Example 2.6, one gets

$$[wt \% O] = 0.0059\% \quad \text{or} \quad 59 \text{ ppm}$$

2.6.2   *Simple deoxidation by manganese, silicon and aluminium*

A thermodynamic treatment of simple deoxidation refers to the ternary system Fe–X–O, where X = Mn, Si, Al, ..., and the individual element (X) forms a simple oxide. At a given temperature, the deoxidizing power of the element thus depends upon its activity in the metal phase and the stability of the oxide in the slag phase. The equilibrium dissolved oxygen contents at 1873 K for different elements are compared in Figure 2.11. The details of the calculation procedure of the simple deoxidation by manganese, silicon and aluminium, which are most commonly used in steelmaking, are given below.

Deoxidation by manganese

Both iron and manganese in the metal phase, as well as FeO and MnO in the slag phase, form a nearly ideal solution. Manganese alone is not used as a deoxidizer,

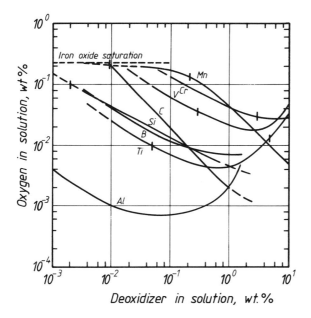

Figure 2.11 Deoxidation equilibria for different deoxidizers in liquid Fe–X–O alloys. (*Source*: Turkdogan, E. T. (1975) In *BOF Steelmaking*, Ch. 4, Vol. 2 (ed. R. D. Pehlke, W. F. Porter, R. F. Urban and J. M. Graines), ISS AIME: New York, pp. 1–190

except in rimming steels. It may, however, significantly modify the deoxidation equilibria when added in combination with silicon and/or aluminium, i.e. complex deoxidation (see section 2.6.3).

The basic reactions taking place during deoxidation by manganese are

$$\{\mathrm{Fe}\} + [\mathrm{O}] = \mathrm{FeO(s)} \quad \text{or} \quad (\mathrm{l}) \tag{2.202}$$

$$[\mathrm{Mn}] + [\mathrm{O}] = \mathrm{MnO(s)} \quad \text{or} \quad (\mathrm{l}) \tag{2.203}$$

where (s) and (l) refer to the solid or liquid reference state, respectively, and are used to calculate the respective equilibrium constants

$$K_{\mathrm{Fe}} = \frac{a_{\mathrm{FeO}}}{h_{\mathrm{O}} a_{\mathrm{Fe}}}$$

Since iron is the solvent $a_{\mathrm{Fe}} \simeq 1$. Hence

$$K_{\mathrm{Fe}} = \frac{a_{\mathrm{FeO}}}{h_{\mathrm{O}}} \tag{2.204}$$

and

$$K_{\mathrm{Mn}} = \frac{a_{\mathrm{MnO}}}{h_{\mathrm{Mn}} h_{\mathrm{O}}} \tag{2.205}$$

Since the FeO–MnO slag behaves almost ideally

$$a_{FeO} = X_{FeO} \quad \text{and} \quad a_{MnO} = X_{MnO} \tag{2.206}$$

From Equations (2.202)–(2.206)

$$K_{Mn} = \left(\frac{X_{MnO}}{X_{FeO}}\right)\frac{K_{Fe}}{h_{Mn}} \tag{2.207}$$

For a slag consisting of FeO and MnO only:

$$X_{FeO} + X_{MnO} = 1 \tag{2.208}$$

From Equations (2.207) and (2.208)

$$X_{MnO} = \frac{K_{Mn}h_{Mn}}{K_{Fe} + K_{Mn}h_{Mn}} \tag{2.209}$$

From Equations (2.205), (2.206) and (2.209), and substituting for X_{MnO},

$$h_O = \frac{1}{(K_{Fe} + K_{Mn}h_{Mn})} \tag{2.210}$$

$$h_{Mn} = f_{Mn}[WMn]$$

and

$$h_O = f_O[WO] \tag{2.211}$$

where $[WO]$ and $[WMn]$ are the wt % of dissolved oxygen and manganese at equilibrium. Let the wt % of dissolved carbon, which is invariably present in steel, be $[WC]$. The influence of carbon on f_O and f_{Mn} can be calculated from

$$\lg f_{Mn} = (e_{Mn}^{Mn}[WMn] + e_{Mn}^{C}[WC] + e_{Mn}^{O}[WO]) \tag{2.212}$$

and

$$\lg f_O = e_O^{O}[WO] + e_O^{C}[WC] + e_O^{Mn}[WMn] \tag{2.213}$$

The equilibrium oxygen content can thus be calculated from

$$[WO] = \frac{1}{f_O(K_{Fe} + K_{Mn}h_{Mn})} \tag{2.214}$$

Since the equations involving f_O and f_M are themselves non-linear and contain the $[WO]$ term, an iterative calculation scheme can be set up as follows. For a fixed temperature and known weight percentages of dissolved carbon and manganese ($[WC]$, $[WMn]$), the $[WO]$ values can be calculated using Equations (2.211)–(2.214). During the initial stages of calculation the $e_{Mn}^{O}[WO]$ and $e_O^{O}[WO]$ terms are assumed to be zero. After calculating $[WO]$ from Equation (2.214), successive iteration is performed to re-evaluate f_{Mn} and f_O until the successive values of $[WO]$ converge with less than 0.5% error.

Deoxidation by silicon

Silicon, upon alloying with liquid iron, generates large amounts of heat, i.e. it exhibits exothermic dissolution. The activity of silicon in a liquid Fe–Si alloy thus shows a strong negative deviation from ideal behaviour. Silicon has a great affinity towards oxygen and is often used to control dissolved oxygen in steel (see Figure 2.11 for [Si]–[O] equilibrium and also Example 2.8).

Example 2.8 **Oxygen in equilibrium with silicon dissolved in steel**

Liquid iron at 1873 K is kept in a silica (β-cristobalite SiO_2) crucible and has 0.5 wt % dissolved silicon. It is required to calculate the equilibrium dissolved oxygen content in steel. Given that

$$\{Si\} + (O_2)_g = \langle SiO_2 \rangle \qquad \Delta G_1^o = -946\,348 + 197.643T \text{ J/mol}$$
$$[Si] \qquad\qquad = \{Si\} \qquad\qquad \Delta G_2^o = \quad 131\,503 + \quad 17.238T \text{ J/mol}$$
$$\qquad\qquad + 2[O] = (O_2)_g \qquad \Delta G_3^o = \quad 234\,304 + \quad\; 5.774T \text{ J/mol}$$

$$[Si] + 2[O] = \langle SiO_2 \rangle \qquad \Delta G_4^o = -580\,541 + 220.655T \text{ J/mol}$$

The equilibrium constant for the above reaction is

$$K_{SiO_2} = \frac{a_{SiO_2}}{[h_{Si}][h_O]^2} \qquad (a_{SiO_2} = 1 \text{ for solid silica})$$

At 1873 K

$$\Delta G_4^o = -167\,254.19 \text{ J/mol}$$
$$= -RT \ln K_{SiO_2}$$

hence

$$\lg K_{SiO_2} = 4.664$$
$$= \lg a_{SiO_2} - \lg[h_{Si}] - 2 \lg[h_O]$$

Now

$$\lg[h_{Si}] = \lg f_{Si} + \lg[\text{wt \% Si}]$$

and

$$\lg[h_O] = \lg f_O + \lg[\text{wt \% O}]$$

The values of the interaction parameters are

$$e_{Si}^{Si} = 0.11 \quad e_{Si}^{O} = -0.25 \quad e_O^{Si} = -0.14 \quad e_O^{O} = -0.2$$

Thus

$$\lg f_{Si} = e_{Si}^{Si}[\text{wt \% Si}] + e_{Si}^{O}[\text{wt \% O}]$$

$$= 0.11[0.5] - 0.25[\text{wt \% O}]$$

$$\lg f_O = e_O^{O}[\text{wt \% O}] + e_O^{Si}[\text{wt \% Si}]$$

$$= -0.20[\text{wt \% O}] - 0.14[0.5]$$

On substituting the requisite values of f_{Si} and f_O and also $a_{SiO_2} = 1$ for pure silica:

$$4.664 = -\lg f_{Si} - \lg[\text{wt \% Si}] - 2 \lg f_O - 2 \lg[\text{wt \% O}]$$

$$= -0.11[0.5] + 0.25[\text{wt \% O}]$$

$$- \lg[0.5] + 2[0.20][\text{wt \% O}]$$

$$+ 2 \times 0.14 \times 0.5 - 2 \lg[\text{wt \% O}]$$

$$= 0.386 + 0.65 \times [\text{wt \% O}] - 2 \lg[\text{wt \% O}]$$

Or

$$2 \lg[\text{wt \% O}] - 0.65[\text{wt \% O}] + 4.278 = 0$$

The above non-linear equation can be solved by a suitable numerical method. As a first approximation, if the term $-0.65[\text{wt \% O}]$ is dropped, then

$$[\text{wt \% O}] = 0.0073\% \quad \text{or} \quad 73 \text{ ppm}$$

If, in addition to silicon, other alloying elements like carbon, manganese, phosphorus and sulphur are present in steel, then f_O and f_{Si} have to be modified as

$$\lg f_{Si} = e_{Si}^{Si}[\text{wt \% Si}] + e_{Si}^{O}[\text{wt \% O}]$$

$$+ e_{Si}^{C}[\text{wt \% C}] + e_{Si}^{Mn}[\text{wt \% Mn}]$$

$$+ e_{Si}^{S}[\text{wt \% S}] + e_{Si}^{P}[\text{wt \% P}]$$

$$\lg f_O = e_O^{O}[\text{wt \% O}] + e_O^{Si}[\text{wt \% Si}] + e_O^{C}[\text{wt \% C}] + e_O^{Mn}[\text{wt \% Mn}]$$

$$+ e_O^{S}[\text{wt \% S}] + e_O^{P}[\text{wt \% P}]$$

The following approach can be used for the calculation of deoxidation equilibria with silicon, when the slag contains FeO and SiO_2.

The fundamental reactions taking place during deoxidation with silicon are

$$\{Fe\} + [O] = FeO(s) \quad \text{or} \quad (1) \tag{2.215}$$

$$[Si] + 2[O] = SiO_2(s) \quad \text{or} \quad (1) \tag{2.216}$$

The equilibrium constants for the above reactions are

$$K_{Fe} = \frac{a_{FeO}}{h_O a_{Fe}}$$

Since iron is the solvent $a_{Fe} \simeq 1$. Hence

$$K_{Fe} = \frac{a_{FeO}}{h_O} \tag{2.217}$$

and

$$K_{Si} = \frac{a_{SiO_2}}{h_{Si} h_O^2} \tag{2.218}$$

The $FeO-SiO_2$ phase equilibrium diagram is shown in Figure 2.12. The Kapoor–Frohberg model described earlier in section 2.4.3 can be used to estimate the activities of FeO and SiO_2. For a given slag composition and hence for a known activity of FeO, a_{FeO} in Equation (2.217), the value of h_O is calculated. Using the corresponding value of the activity of SiO_2, a_{SiO_2} in Equation (2.218), h_{Si} is calculated. Let the wt % of oxygen, silicon and carbon be [WO], [WSi] and [WC], respectively. The activities of silicon and oxygen can be calculated from

$$\lg h_{Si} = \lg[WSi] + e_{Si}^{Si}[WSi] + e_{Si}^{C}[WC] + e_{Si}^{O}[WO] \tag{2.219}$$

and

$$\lg h_O = \lg[WO] + e_O^{Si}[WSi] + e_O^{C}[WC] + e_O^{O}[WO] \tag{2.220}$$

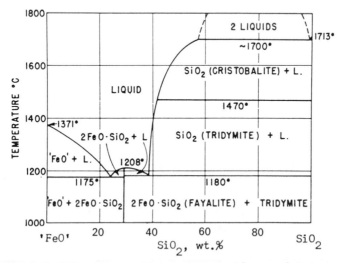

Figure 2.12 $FeO-SiO_2$ phase equilibrium diagram. (*Source*: Schuhmann, R. and Ensio, P. J. (1951) *Trans. Metall. Soc. AIME*, **191**, 401–11; Michal, E. J. and Schuhmann, R. (1952) *Trans. Metall. Soc. AIME*, **194**, 708–23.)

For a fixed $[WC]$, it is possible to solve Equations (2.219) and (2.220) for $[WSi]$ and $[WO]$ simultaneously.

Deoxidation by aluminium

The effect of aluminium on the dissolved oxygen content of steel, assuming that the deoxidation product is pure solid Al_2O_3, is shown in Figure 2.11.

The basic reactions during deoxidation by aluminium are

$$\{Fe\} + [O] = FeO(s) \text{ or } (1) \tag{2.221}$$

$$2[Al] + 3[O] = Al_2O_3(s) \tag{2.222}$$

The equilibrium constants for the above reactions are

$$K_{Fe} = \frac{a_{FeO}}{h_O a_{Fe}} \tag{2.223}$$

and

$$K_{Al} = \frac{a_{Al_2O_3}}{h_{Al}^2 h_O^3} \tag{2.224}$$

At a given temperature and for known slag activity data the value of h_O can be calculated. From the activity of Al_2O_3 and Equation (2.224), the value of h_{Al} can be calculated. Now when wt % of Al is $[WAl]$, then

$$\lg h_{Al} = \lg[WAl] + e_{Al}^{Al}[WAl] + e_{Al}^C[WC] + e_{Al}^O[WO] \tag{2.225}$$

and

$$\lg h_O = \lg[WO] + e_O^{Al}[WAl] + e_O^C[WC] + e_O^O[WO] \tag{2.226}$$

For a fixed value of $[WC]$, it is possible to solve for $[WAl]$ and $[WO]$ in the simultaneous Equations (2.225) and (2.226).

2.6.3 *Complex deoxidation*

A combination of deoxidizers like (Si + Mn), or (Al + Si) or (Al + Si + Mn), or (Ca + Al), or (Ca + Al + Si), etc., can give lower oxygen contents than when the elements are added individually. This is due to the formation of not only complex but also thermodynamically more stable deoxidation products. The objective is to form a liquid deoxidation product which can float up easily (so as to reduce the inclusion content). In performing complex deoxidation calculations, the activity–composition data for the slag are obtained from the Kapoor–Frohberg model (described in section 2.4.3). The calculation procedure for (Mn + Si) deoxidation, without the incorporation of FeO, is described below.

The reactions taking place during simultaneous deoxidation by $(Mn + Si)$ are

$$[Mn] + [O] = MnO(s) \text{ or } (1) \tag{2.227}$$

and

$$[Si] + 2[O] = SiO_2(s) \text{ or } (1) \tag{2.228}$$

The equilibrium constants for the above reactions are

$$K_{Mn} = \frac{a_{MnO}}{h_{Mn}h_O} \tag{2.229}$$

and

$$K_{Si} = \frac{a_{SiO_2}}{h_{Si}h_O^2} \tag{2.230}$$

The MnO–SiO_2 phase diagram is shown in Figure 2.13 and the liquid range is given in Table 2.4.

We know that

$$h_{Mn} = f_{Mn}[WMn] \tag{2.231}$$

$$h_{Si} = f_{Si}[WSi] \tag{2.232}$$

and

$$h_O = f_O[WO] \tag{2.233}$$

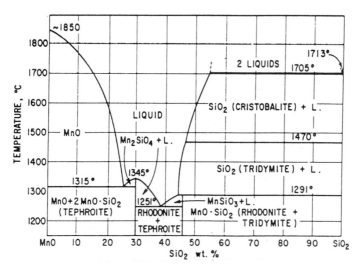

Figure 2.13 MnO–SiO$_2$ phase equilibrium diagram. (*Source*: Glasser, F. P. (1958) *Am. J. Sci.*, **256**, Ser-5, 318–412.)

Table 2.4　Liquid range in M_xO_y–N_pO_q type slag.

System	Temperature (°C)	Liquid range (wt % N_pO_q)	
		From	To
MnO–SiO_2	1500	22.73	48.18
	1650	18.18	52.72
CaO–SiO_2	1500	44.29	65.71
CaO–Al_2O_3	1500	44.35	54.78

Also

$$\lg f_{Si} = e_{Si}^{Si}[WSi] + e_{Si}^{Mn}[WMn] + e_{Si}^{C}[WC] + e_{Si}^{O}[WO] \qquad (2.234)$$

$$\lg f_{O} = e_{O}^{Si}[WSi] + e_{O}^{Mn}[WMn] + e_{O}^{C}[WC] + e_{O}^{O}[WO] \qquad (2.235)$$

and

$$\lg f_{Mn} = e_{Mn}^{Si}[WSi] + e_{Mn}^{Mn}[WMn] + e_{Mn}^{C}[WC] + e_{Mn}^{O}[WO] \qquad (2.236)$$

For a fixed value of $[WMn]$, $[WSi]$, $[WC]$, temperature and known slag activity data (i.e. from the Kapoor–Frohberg model described in section 2.4.3) the

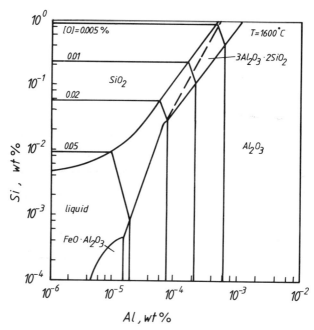

Figure 2.14　Fe–Al–Si–O equilibria at 1873 K. (*Source*: Schürmann, E. and Bannenberg, N. (1984) *Arch. Eisenhüttenwes.*, **55**, 455–62.)

values of f_{Mn}, f_{Si} and f_O can be calculated using Equations (2.234)–(2.236). Initially the terms $e_i^O[WO]$, where $i = $ Mn, Si and O, are assumed to be zero. Using Equations (2.231) and (2.232), the values of h_{Mn} and h_{Si} can be found subsequently.

From Equations (2.229), (2.233) and (2.236)

$$[WO] = \frac{a_{MnO}}{h_{Mn} f_O K_{Mn}} \tag{2.237}$$

The value of $[WO]$ obtained in Equation (2.237) is used iteratively to re-estimate f_{Mn}, f_{Si} and f_O until the two consecutive values agree with less than 0.5% error.

The Al–Si–O and Al–Si–Mn–O equilibria are presented in Figures 2.14 and 2.15. The lowest oxygen contents can be obtained with Al–Si–Mn but extreme care is required to adjust the steel composition such that the deoxidation product is liquid.

It should be noted that the reaction of deoxidizers with extraneous sources of oxygen, such as top slag rich in FeO, atmospheric oxygen (during tapping and teeming), refractory lining, etc., always introduces an uncertainty or correction factor in the amount of deoxidizer to be added (see Chapter 7).

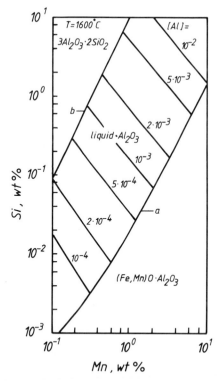

Figure 2.15   Fe–Mn–Al–Si–O equilibria at 1873 K. (*Source*: Schürmann, E. and Bannenberg, N. (1984) *Arch. Eisenhüttenwes.*, **55**, 455–62.)

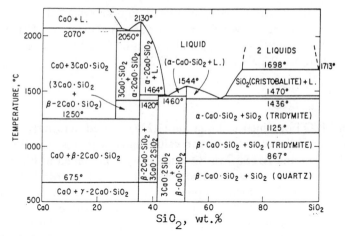

Figure 2.16 CaO–SiO₂ phase equilibrium diagram. (*Source*: Welch, J. H. and Gott, W. (1959) *J. Am. Ceram. Soc.*, **42**, 11–15.)

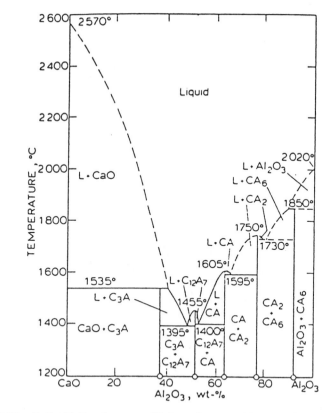

Figure 2.17 CaO–Al₂O₃ phase equilibrium diagram. (*Source*: Salter, W. J. M. and Pickering, F. B. (1969) *J. Iron Steel Inst.*, **207**, 992–1002.)

The control of deoxidation with calcium (namely, Ca–O, Ca–Al–O, Ca–Al–Si, Ca–Mn–Al–O, etc.) is even more complex because of the lack of reliable thermodynamic data on the dissolution of calcium in liquid steel. The liquid range in the $CaO–Al_2O_3$ phase diagram (as also indicated in Table 2.4) is small. Calcium (similar to magnesium) also forms a gaseous phase but the recovery of calcium and the associated reaction mechanisms with oxygen, silicon, manganese and aluminium are very difficult to predict. It is common practice to deoxidize the steel first by Al or (Al + Si). Calcium-based agents (in the form of a cored wire containing Ca–Si) are then added to modify the inclusion morphology. The liquid range in the $CaO–SiO_2$ phase diagram (Figure 2.16 and Table 2.4) is wider than in the $CaO–Al_2O_3$ phase diagram (Figure 2.17). For example, calcium in excess of 15 ppm converts the galaxy-shaped alumina inclusions to globular calcium aluminate. The MnS inclusions do not appear in 0.005 wt % sulphur steels containing calcium in excess of 80 ppm (see Chapter 7).

2.7 Compilation of thermodynamic data for steelmaking

Thermodynamic data form the basis of all the calculations in steelmaking and refining. Researchers are engaged not only in experimentation (using more accurate methods), but also in reassessing, updating and recompiling reliable thermodynamic data pertaining to liquid iron alloys. Once the experimental data are reported in the literature, the accuracy of the experimental methods used by different workers can be compared. Thermodynamic models and numerical and statistical methods are employed for evaluation. As an example of the procedure of evaluation of free energy and interaction parameter data in Fe–C–O, Fe–Cr–O, Fe–Al–O, Fe–Zr–O, Fe–V–O and Fe–Ti–O alloys, reference may be made to Deo [49] and Ghosh and Murthy [50]. The new recommended values of equilibrium constant and interaction parameter data pertaining to all these alloys are included in Tables 2.1 and 2.2.

The compilations of the phase diagrams and the activity–composition diagrams relevant to steelmaking slags are also available (see [1–10] in Chapter 1). As a starting point, however, since components other than CaO, FeO and SiO_2 add up to less than 20% by mass, the binary $CaO–SiO_2$ and $FeO–SiO_2$, $CaO–FeO$ and the quasi-ternary $CaO'–FeO*–SiO_2'$ systems are important for the study of steelmaking slags. In a quasi-ternary system the percentages of CaO, FeO and SiO_2 are recalculated assuming that wt % CaO + wt % FeO + wt % SiO_2 = 100%, and the new values thus obtained are indicated by CaO', SiO_2' and FeO*. The FeO* in the slag represents total iron in slag expressed as FeO. The Fe–FeO–Fe_2O_3 equilibrium is written as

$$Fe_{(1)} + Fe_2O_{3(1)} = FeO_{(1)}$$

The total iron content (Fe_t) of slag is reported according to the equation

$$FeO + Fe_2O_3 = FeO_n$$

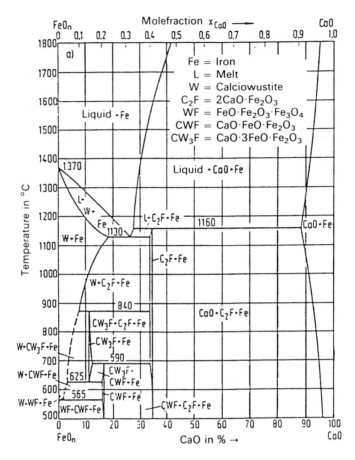

Figure 2.18   CaO–FeO phase equilibrium diagram. (*Source*: Schürmann, E. and Kraume, G. (1976) *Arch. Eisenhüttenwes.*, **47**, 327–38.)

and

$$FeO_n = k(Fe_t)$$

The approximate values of k and n as a function of wt % FeO_n, are as follows:

FeO_n (wt %)	k	n
15	1.308	1.075
20	1.315	1.100
25	1.322	1.125

The phase diagram of FeO–SiO$_2$ is shown in Figure 2.12. The phase diagrams of CaO–SiO$_2$, CaO–FeO and CaO–FeO*–SiO$_2$ in equilibrium with iron are shown in Figures 2.16, 2.18 and 2.19, respectively. The activity of FeO in CaO–FeO–SiO$_2$ slags at 1873 K is shown in Figure 2.6. The activities of SiO$_2$ and CaO are shown in Figures 2.20 and 2.21, respectively.

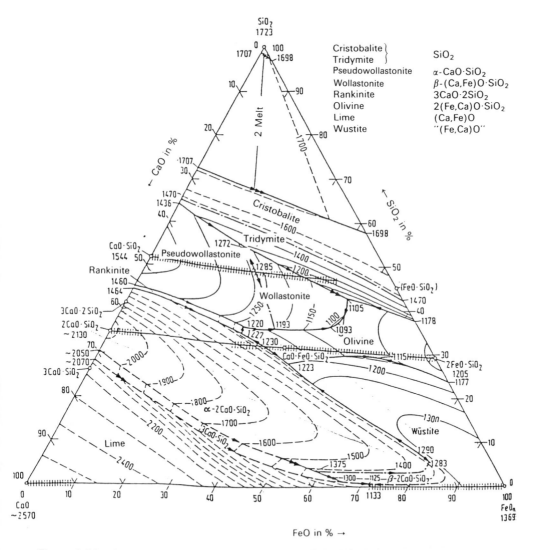

Figure 2.19   CaO–FeO–SiO$_2$ phase equilibrium diagram. (*Source*: Osborn, E. F. and Muan, A. (1960) *Phase equilibrium diagrams of oxide systems, Ceramic Foundation,* E. Orton Jr.: Columbus, OH.)

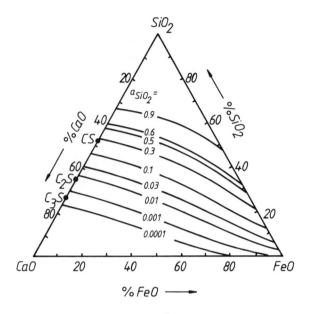

Figure 2.20   Activity of CaO in CaO–FeO–SiO$_2$ slags at 1873 K (calculated from Kapoor–Frohberg model in section 2.4.3).

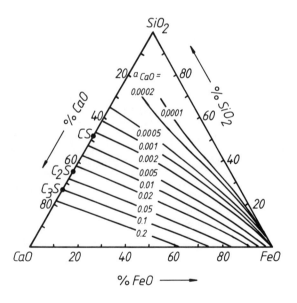

Figure 2.21   Activity of SiO$_2$ in CaO–FeO–SiO$_2$ slags at 1873 K (calculated from Kapoor–Frohberg model in section 2.4.3).

References

1. De Boer, F. R., Boom, R., Mattens, W. C. M., Miedema, A. R. and Niessen, A. K. (1988) *Cohesion in Metals – Transition Metal Alloys, Cohesion and Structure Series*, North-Holland: Amsterdam.
2. Turkdogan, E. T. (1975) 'Physical chemistry of oxygen steel making, thermochemistry and thermodynamics', In *BOF Steelmaking* (ed. R. D. Pehlke, W. F. Porter, R. F. Urban and J. M. Gaines), Chapter 4, Vol. 2, New York: ISS AIME pp. 1–190.
3. Chipman, J. (1970) 'Thermodynamics of liquid Fe–C solutions', *Metall. Trans.*, **1**, 2163–8.
4. Flood, H. and Grjotheim, K. (1952) 'Thermodynamic calculation of slag equilibria', *J. Iron Steel Inst.*, **171**, 64–70.
5. Gaye, H. and Grosjean, J.-C. (1982) 'Metallurgical reactions in the LBE process', *Proc. 65th National Open Hearth and Basic Oxygen Steelmaking Conf.*, Pittsburgh, pp. 202–10.
6. Guo, D. (1984) 'A new model predicting phosphorus equilibrium and its applications', *Arch. Eisenhüttenwes.*, **55**, 183–8.
7. Balajiva, K., Quarrel, A. G. and Vajragupta, P. (1946) 'A laboratory investigation of the phosphorus reaction in the basic steelmaking process', *J. Iron Steel Inst.*, **153**, 115–50.
8. Turkdogan, E. T. (1980) *Physical Chemistry of High Temperature Technology*, Academic Press: New York.
9. Kor, G. J. W. (1977) 'Effect of fluorspar and other fluxes on slag–metal equilibria involving phosphorus and sulfur', *Metall. Trans.*, **8B**, 107–13.
10. Jeffes, J. H. E. (1980) Private communication with B.D. (London, Imperial College).
11. Healy, G. W. (1970) 'A new look at phosphorus distribution', *J. Iron Steel Inst.*, **208**, 664–8.
12. Turkdogan, E. T. (1984) 'Physicochemical aspects of reactions in ironmaking and steelmaking processes', *Trans. Iron Steel Inst. Jpn*, **24**, 591–611.
13. Bauer, K.-H., Bock, M., Wagner, H. and Bannenberg, N. (1986) 'Zusammenhaug zwischen Roheisenzu-sammensetzung, Kalksatz und Konverterhalf-barkeit beim Arbeiten nach dem LD-b2w, kombinierten Blasverfahren' *Stahl und Eisen*, **106**, 389–97.
14. Endell, K. and Hellbrugge, J. (1942) 'Über den Einfluss des Iron Raders und der Wertigkeit der Rationen auf die electrische Leitfahigkeit von Silicatschmelzen', *Naturwissenschaften*, **30**, 421–2.
15. Flood, H. and Förland, T. (1947) 'The acidic and basic properties of oxides', *Acta Chem. Scand.*, **1**, 592–604 and 781–9.
16. Bockris, J. O' M., Mackenzie, J. D. and Kitchner, J. A. (1955) 'Viscous flow in silica and binary liquid silicates', *Trans. Faraday Soc.*, **51**, 1734–48.
17. Fincham, C. J. B. and Richardson, F. D. (1954) 'The behaviour of sulphur in silicate and aluminate melts', *Proc. R. Soc.*, **A223**, 40–62.
18. Derge, G. (1965) 'Highly basic slags', In *Steelmaking: The Chipman Conference*, MIT Press: Cambridge, MA, pp. 88–94.
19. Masson, C. R. (1973) 'Thermodynamics and constitution of silicate slags', In *Chemical Metallurgy of Iron and Steel*, Iron and Steel Institute: London, pp. 3–11.
20. Toop, G. W. and Samis, C. S. (1962) 'Activities of ions in silicate melts', *Trans. Metall. Soc. AIME* **224**, 878–87.
21. Ban-ya, S. and Shim, J. D. (1982) 'Application of the regular solution model for the equilibrium of distribution of oxygen between liquid iron and steelmaking slags', *Can. Metall. Q.*, **21**, 319–28.

22. Frohberg, M. G. and Kapoor, M. L. (1971) 'Die Anwendung eines neuen Basizitäts maßes auf metallurgische Reaktionen' *Stahl und Eisen*, **91**, 182–8.
23. Flood, H. and Knapp, W. J. (1963) 'Acid–base equilibria in the system $PbO-SiO_2$', *J. Am. Ceram. Soc.*, **46**, 61–5.
24. Gaskell, D. R. (1981) 'Thermodynamic models of liquid silicates', *Can. Metall. Q.*, **20**, 3–19.
25. Lumsden, J. (1961) 'The thermodynamics of liquid iron silicates' In *Physical Chemistry of Process Metallurgy*, Pt 1, Interscience: New York, pp. 165–205.
26. Sommerville, I. D., Ivanchev, I. and Bell, H. B. (1973) 'Activity in melts containing FeO, MnO, Al_2O_3 and SiO_2', In *Chemical Metallurgy of Iron and Steel*, Iron and Steel Institute: London, pp. 23–5.
27. Masson, C. R. (1965) 'An approach to the problem of ionic distribution in silicate slags', *Proc. R. Soc.*, **A287**, 201–21.
28. Masson, C. R., Smith, I. B. and Whiteway, S. G. (1970) 'Molecular size distribution in multichain polymers: Applications of polymer theory', *Can. J. Chem.*, **48**, 1456–64.
29. Abraham, K. P. and Richardson, F. D. (1961) 'The mixing of silicates in molten slags' In *Physical Chemistry of Process Metallurgy*, Pt. 1, Interscience: New York, pp. 263–276.
30. Kapoor, M. L. and Frohberg, M. G. (1973) 'Theoretical treatment of activities in silicate melts', In *Chemical Metallurgy of Iron and Steel*, Iron and Steel Institute: London, pp. 17–22.
31. Frohberg, M. G. and Kapoor, M. L. (1970) 'Die elektrolytische Dissoziation flüssiger Schlacken und ihre, Bedeutung für metallurgische Reaktionen' *Arch. Eisenhüttenwes.*, **41**, 209–12.
32. Yokokawa, T. and Niwa, K. (1969) 'Free energy of solution in binary silicate melts', *Trans. Jpn Inst. Met.* **10**, 3–7.
33. Kapoor, M. L. and Frohberg, M. G. (1977) 'Activity of lead oxide in the system sodium oxide lead oxide–silica', *Metall. Trans.*, **8B**, 15–18.
34. Yokowa, T. and Niwa, K. (1969) 'Free energy and basicity of molten silicate solutions', *Trans. Jpn Inst. Met.*, **10**, 81–4.
35. Kapoor, M. L., Mehrotra, G. M. and Frohberg, M. G. (1975) 'Statistical thermodynamic approach to chain polymers in binary silicate melts', *Proc. Aust. Inst. Miner. Metall.*, **254**, 11–17.
36. Kapoor, M. L., Mehrotra, G. M. and Frohberg, M. G. (1974) 'Zusammenhang zwischen den thermodynamischen Größen und der Struktur flüssiger Silicatsysteme' *Arch. Eisenhüttenwes.* **45**, 213–18.
37. Kapoor, M. L., Mehrotra, G. M. and Frohberg, M. G. (1974) 'Die Berechnung thermodynamischer Größen und Struktureller Eigenschaften flüssiger binärer Silicatsysteme' *Arch. Eisenhüttenwes.*, **45**, 663–9.
38. Gaye, H. and Welfringer, J. (1984) 'Modelling of the thermodynamic properties of complex metallurgical slags', In *Second Int. Symp. on Metallurgical Slags and Fluxes* (ed. H. A. Fine and D. R. Gaskell), Metall. Soc. AIME: Warrendale, PA, pp. 357–75.
39. Boom, R., Deo, B., van der Knoop, W., Mensonides, F. and van Unen, G. (1989) 'The application of models for slag formation in basic oxygen steelmaking', *Proc. 3rd Int. Conf. on Molten Slags and Fluxes, Glasgow, 27–29 June 1988*, The Institute of Metals: London, pp. 273–6.
40. Richardson, F. D. (1974) *Physical Chemistry of Melts in Metallurgy*, Vol. 2, Academic Press: London.
41. Bell, H. B. (1989) 'Some applications of slag chemistry to metal refining', *Proc. 3rd Int. Conf. on Molten Slags and Fluxes, Glasgow, 27–29 June 1988*, The Institute of Metals: London, pp. 277–82.

42. Chen, B., Reddy, R. G. and Blander, M. (1989) 'Sulphide capacities of $CaO-FeO-SiO_2$ slags', *Proc. 3rd Int. Conf. on Molten Slags and Fluxes, Glasgow, 27–29 June 1988*, The Institute of Metals: London, pp. 270–2.
43. Wagner, C. (1975) 'The concept of basicity of slags', *Metall. Trans.*, **6B**, 405–9.
44. Duffy, J. A. and Ingram, M. D. (1976) 'An interpretation of glass chemistry in terms of the optical basicity concept', *J. Non-cryst. Solids*, **21**, 373–410.
45. Duffy, J. A. and Ingram, M. D. (1975) 'Influence of electronegativity on the Lewis basicity and solvent properties of molten oxyanion salts and glasses', *J. Inorg. Nucl. Chem.*, **37**, 1203–6.
46. Sosinsky, D. J. and Sommerville, I. D. (1986) 'The composition and temperature dependence of the sulfide capacity of metallurgical slags', *Metall. Trans.*, **17B**, 331–7.
47. Bergman, A. and Gustafsson, A. (1988) 'On the relation between optical basicity and phosphorus capacity of complex slags', *Steel Res.*, **59**, 281–8.
48. Steinmetz, E. and Lindenberg, H.-U. (1970) 'Kinetik der Reaktionen zwischen Kohlenstoff und Sauerstoff bei der Stahlherstellung' *Stahl und Eisen*, **90**, 1517–25.
49. Deo, B. (1988) 'Assessment of equilibrium constants and interaction parameters in liquid Fe–C–O alloys', *Trans. Indian Inst. Met.*, **41**, 237–46.
50. Ghosh, A. and Murthy, G. V. R. (1986) 'An assessment of thermodynamic parameters for deoxidation of molten iron by Cr, V, Al, Zr and Ti', *Trans. Iron Steel Inst. Jpn*, **26**, 629–37.

CHAPTER

3 *Kinetic fundamentals*

3.1 Kinetic steps

The reactions in steelmaking involve the simultaneous reactions of many components at the interfaces of several phases, i.e. a heterogeneous, multicomponent, multiphase system at a high temperature (see Figure 1.1). The feasibility of a chemical reaction and the equilibrium compositions of the coexisting phases can be predicted by thermodynamics (discussed in Chapter 2). A knowledge of the reaction rates or kinetics is, however, required to predict the changes occurring in a particular process, as a function of time, while it is approaching equilibrium.

The overall rate of a particular reaction, at any given moment, depends upon the properties of the phases involved in the reaction and the kinetic steps in going from the initial to the final state. In principle, a reaction which involves more than two phases, for instance the oxidation of iron by oxygen gas in a three-phase gas–metal–slag system

$$\{Fe\} + \tfrac{1}{2}(O_2)_g = (FeO)$$

is a combination of two separate reactions involving two phases at a time:

1. Gas–metal reaction: dissolution of oxygen in steel; dissolved oxygen is represented as $[O]$, $(O_2)_g = \tfrac{1}{2}[O]$.
2. Slag–metal reaction: formation of FeO, $\{Fe\} + [O] = (FeO)$.

The reaction takes place in two steps because of the basic physical limitation that only two phases, at a time, can meet at a plane whereas three phases can intersect at a line only. For the sake of simplicity, consider a simple reaction

$$A(1) + B(1) = AB(2) \tag{3.1}$$

involving only two components and two immiscible phases (i.e. metal–gas, slag–gas or slag–metal) where A and B are soluble in phase (1), AB is soluble in phase (2) only and phases (1) and (2) are immiscible in each other. The steps involved in the

106

above reaction taking place at the interface of phases (1) and (2) are as follows:

1. The mass transfer of A from the bulk of phase (1) to the interface of phases (1) and (2).
2. The mass transfer of B from the bulk of phase (1) to the interface of phases (1) and (2).
3. The chemical reaction of A and B at the interface to produce AB.
4. The mass transfer of the product AB away from the interface into the bulk of phase (2).
5. When two immiscible phases are brought into contact it is nearly always found that the concentration of a component is greater at the interface than in the bulk. This tendency for accumulation to take place at a surface is called adsorption. Desorption is the inverse of adsorption.

Each one of the above steps is known as a kinetic step. At any given moment the slowest kinetic step would control the rate of the reaction and is termed the 'rate-controlling step' or the 'rate-limiting step'. In a real system, with several phases and interfaces (as shown in Figure 1.1), the reactions at some interfaces may be controlled only by mass transfer, while at the other interfaces they may be controlled by chemical reaction, or by desorption/adsorption, or by a combination of more than one step, namely chemical reaction plus mass transfer, adsorption plus mass transfer, etc. Therefore, a good understanding of metallurgical kinetics requires knowledge of the mechanisms of mass transfer, chemical reaction and adsorption/desorption.

3.2 Mechanisms of mass transfer

Mass transfer in a fluid can occur by several mechanisms:

1. Molecular or atomic diffusion Molecular or atomic diffusion in a solid, liquid or gas occurs as a result of the presence of a concentration gradient. According to Fick's first law, the diffusive flux J_{D_m} (number of moles per unit time per unit area) is proportional to the concentration gradient, $\partial c / \partial x$:

$$J_{D_m} = -D_m(\partial c / \partial x) \tag{3.2}$$

where D_m is the atomic (or molecular) diffusion coefficient.

2. Laminar flow of the fluid The laminar flow or convection in a liquid directly transports the dissolved or entrained material from one place to another. Fick's second law describes the equivalence between local convective diffusion and molecular diffusion. Let u_x, u_y and u_z be the respective fluid velocities in the x, y and *z* directions. Then, similar to Equations (1.18) and (1.19):

$$\frac{\partial c}{\partial t} + \left(u_x \frac{\partial c}{\partial x} + u_y \frac{\partial c}{\partial y} + u_z \frac{\partial c}{\partial z} \right) = D_m \left(\frac{\partial^2 c}{\partial x^2} + \frac{\partial^2 c}{\partial y^2} + \frac{\partial^2 c}{\partial z^2} \right) \tag{3.3}$$

For uniaxial steady-state diffusion:

$$\frac{\partial c}{\partial t} = 0, \quad u_y = u_z = 0, \quad \frac{\partial^2 c}{\partial y^2} = \frac{\partial^2 c}{\partial z^2} = 0 \tag{3.4}$$

and

$$u_x \frac{\partial c}{\partial x} = D_m \frac{\partial^2 c}{\partial x^2} \tag{3.5}$$

3. Eddy or turbulent flow   In the presence of turbulent flow, an additional mass transfer will take place as a result of the eddy packets. Similar to Equation (3.2)

$$J_{D_t} = -D_t \frac{\partial c}{\partial x} \tag{3.6}$$

where D_t is the eddy or turbulent diffusion coefficient. The total flux (J_D) is thus

$$J_D = J_{D_m} + J_{D_t} \tag{3.7}$$

The mathematical treatment of the effect of fluid flow on the mass transfer taking place at the interface of two immiscible phases is greatly simplified by employing mass transfer coefficients in Equations (3.2)–(3.6). Consider, for example, a weakly agitated melt in which a dissolved element in the melt (say, hydrogen in steel) is transferred to the gas phase. We may assume that a thin film forms at the surface of the melt in which the flow is laminar. The solute will diffuse through this film to the gas phase. Figure 3.1 shows the curves of hypothetical concentration versus distance and velocity versus distance in the metal phase. The tangent to the concentration profile intersects the distance axis at δ and is called the thickness of the concentration boundary layer. The tangent to the velocity profile intersects at δ' and is called the velocity boundary layer. Since δ' is greater than δ in weakly agitated melts, the concentration boundary layer determines the mass transfer rate. Both δ' and δ decrease with increasing fluid velocity. When a steady state is reached, δ' and δ become constant. Let $\delta = \partial x$ and $\partial C = (C_m^b - C_m^i)$. Equation (3.2) then becomes

$$J_{D_m} = k_m^D (C_m^b - C_m^i) \tag{3.8}$$

where C_m^i and C_m^b are the interfacial and bulk concentrations (kmol/m^3), respectively, and

$$k_m^D = D_m / \delta \tag{3.9}$$

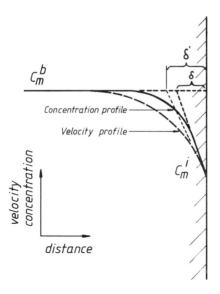

Figure 3.1 Velocity and concentration profiles near a solid–liquid interface; δ is the thickness of the concentration boundary layer and δ' is the thickness of the velocity boundary layer.

where k_m^D is called the diffusion mass transfer coefficient (m/s) in the metal phase. It is obvious that in a stirred liquid the value of k_m^D would depend not only upon the factors affecting the diffusion coefficient D_m, but also upon the physical state of the system affecting δ, such as bath agitation intensity and system geometry. At the interface of immiscible liquids (in the absence of turbulent diffusion), Equation (3.7) reduces to

$$J_D = J_{D_m} = k_m^D(C_m^b - C_m^i)$$ (3.10)

In a general case, it can be assumed that both the diffusion mass transfer and the turbulent mass transfer processes will take place and the overall mass transfer coefficient, k_m, will be given by

$$k_m = A_D k_m^D + (1 - A_D)k_m^E$$ (3.11)

where A_D is the fractional contribution of diffusion mass transfer and k_m^E is the turbulent mass transfer coefficient. It is known from experiments that k_m^E is proportional to the velocity of the fluid element, u_n, normal to the interface. Thus

$$k_m = A_D k_m^D + \alpha(1 - A_D)u_n$$ (3.12)

where α is the constant of proportionality.

3.2.1 *Effect of fluid flow on mass transfer coefficient*

In an agitated bath, the velocity (**u**) of molten metal can be computed from the Navier–Stokes equation

$$\rho\left(\frac{\partial \mathbf{u}}{\partial t} + \mathbf{u}\nabla\mathbf{u}\right) = -\nabla p + \mu_e \nabla^2 \mathbf{u} + \mathbf{F} \tag{3.13}$$

which is similar to Equation (1.19); **F** denotes the body force generated by gas injection or mechanical stirring, electromagnetic stirring, etc., μ_e is the effective viscosity, p is the pressure and ρ is the density. The solution of Equation (3.13) can be considered under three specific situations of fluid flow in an agitated melt [1] as follows:

1. **Fluid motion is dominated by viscous force**   Equation (3.13) reduces to

$$\mu_e \nabla^2 \mathbf{u} = -\mathbf{F} \tag{3.14}$$

In principle, the differential operator ∇ can be replaced by the dimension of length L^{-1} and on integrating Equation (3.14)

$$u \propto (FL^2/\mu_e) \tag{3.15}$$

2. **Fluid motion is dominated by inertia force**   Equation (3.13) reduces to

$$\rho\mathbf{u}\nabla\mathbf{u} = \mathbf{F} \tag{3.16}$$

Similar to Equation (3.15)

$$u \propto (FL/\rho)^{1/2} \tag{3.17}$$

3. **Fluid motion is dominated by turbulent viscous force**   Equation (3.13) reduces to

$$\mu_t \nabla^2 \mathbf{u} = -\mathbf{F} \tag{3.18}$$

According to Boussinesq and Prandtl's hypothesis, the turbulent viscosity μ_t can be expressed by

$$\mu_t = \rho l^2 |\nabla u| \tag{3.19}$$

where l is the mixing length. On substituting Equation (3.19) into Equation (3.18):

$$u = (Fl^3/\rho l^2)^{1/2} = (Fl/\rho)^{1/2} \tag{3.20}$$

The work WK done by a force F during a displacement ds is given by

$$WK = \int F \, ds \tag{3.21}$$

Now, let $u = ds/dt$ and let the mixing power density $\dot{\varepsilon} \equiv dWK/dt$, where $\dot{\varepsilon}$ is expressed in W/ton. Substituting these in Equation (3.21), we obtain on integration

$$\dot{\varepsilon} \propto uF \tag{3.22}$$

The value of F in Equation (3.22) can be substituted in Equations (3.15), (3.17) and (3.20) to yield the following:

(a) for the fluid motion dominated by viscous force:

$$u \propto (L^2 \dot{\varepsilon}/\mu)^{1/2} \tag{3.23}$$

or

$$u \propto \dot{\varepsilon}^{1/2} \tag{3.24}$$

(b) for the fluid motion dominated by the force of inertia:

$$u \propto (L\dot{\varepsilon}/\rho)^{1/3} \tag{3.25}$$

or

$$u \propto \dot{\varepsilon}^{1/3} \tag{3.26}$$

(c) for the fluid motion dominated by turbulent viscous force:

$$u \propto (L^3 \dot{\varepsilon}/\rho L^2)^{1/3} \tag{3.27}$$

or

$$u \propto \dot{\varepsilon}^{1/3} \tag{3.28}$$

In a gas-stirred melt the volume flow rate of gas, Q (m³/s), is approximately proportional to $\dot{\varepsilon}$ and hence, analogous to Equation (3.28),

$$u \propto Q^n \tag{3.29}$$

As stated before, k_m^E is directly proportional to the fluid velocity; hence in the case of a melt vigorously stirred by gas

$$k_m^E \propto \dot{\varepsilon}^n \tag{3.30}$$

or

$$k_m^E \propto Q^n \tag{3.31}$$

where n may range from $1/2$ to $1/3$. For a given experimental situation, let the mass transfer coefficient be expressed in a modified form as the mass transfer capacity coefficient $\mathscr{K}$, i.e.

$$\mathscr{K} = k \frac{A}{V_m} \tag{3.32}$$

where V_m is the volume of metal and A the interfacial slag–metal area. We can express Equation (3.30) as

$$\mathscr{K} \propto \dot{\varepsilon}^n \tag{3.33}$$

The exponent n in Equation (3.33) is less than 0.5 in most cases (Table 3.1), which is also predicted by Equations (3.24), (3.26) and (3.28). For some of the cases in

Table 3.1   Correlation of mass transfer in liquid/liquid systems.

System	Stirring	Reaction	Correlation	Remarks
Slag/steel	Ar gas	Desulphurization	$\mathcal{K} \propto \varepsilon^{0.25}$; $\varepsilon < 60\,\mathrm{W/ton}$ $\mathcal{K} \propto \varepsilon^{2.1}$; $\varepsilon > 60\,\mathrm{W/ton}$	2.5 ton converter $q < 150$ $150 < q < 240$
Water/Hg	N_2 gas	Reduction of quinone	$\mathcal{K} \propto \varepsilon^{0.3-0.4}$	$q < 58$ $\beta = 0.22$
Slag/steel	Ar gas mechanical stirring	Dephosphorization	$\mathcal{K} \propto \varepsilon^{0.60}$	$30 < q < 160$
Slag/steel	Ar gas	$[Cu] \rightarrow (Cu)$	$\mathcal{K} \propto \varepsilon^{0.27}$	$q < 100$
Slag/steel		Desulphurization	$\mathcal{K} \propto \varepsilon^{1.0}$	
Oil/water			$\mathcal{K} \propto \varepsilon^{0.33}$	
Amalgams/aqueous solution		$[In] + 3(Fe^{3+})$ $(In^{3+}) + 3(Fe^{2+})$	$\mathcal{K}_m \propto \varepsilon^{0.33}$ $\mathcal{K}_s \propto \varepsilon^{0.42}$	$q < 10$ $\beta = 0.5$
n-hexane/aqueous solution	N_2 gas	$I_2 + 2OH^- = IO^-$ $+I^- + H_2O$ $3IO^- = IO_3^- + 2I^-$	$\mathcal{K} \propto \varepsilon^{0.72}$	$199 < q < 994$ $\beta = 0.5$
Amalgams/aqueous solution Lead/molten salt			$\mathcal{K} \propto \varepsilon^{0.5}$	$q < 130$ $\beta = 0.5$
Slag/steel	O_2 gas	Dephosphorization	$\mathcal{K} \propto \varepsilon^{0.54}$	$50 < q < 80$
Liquid paraffin/water			$\mathcal{K} \propto \varepsilon^{0.36}$ $\mathcal{K} \propto \varepsilon^{3.0}$	$30 < q < 80$ $80 < q < 200$ $\beta = 0.17$
Tetraline/aqueous solution	Air		$\mathcal{K} \propto \varepsilon^{0.36}$ $\mathcal{K} \propto \varepsilon^{1.0}$	$q < 150$ $150 < q < 650$ $\beta = 0.1$

Notes: $q \propto \varepsilon$ is assumed (q, stirring gas flow rate in litres/min/ton); $\mathcal{K}$, capacity coefficient of mass transfer; ε, mixing power density; β, fraction of slag.
Source: Asai, S., Kawachi, M. and Muchi, I. (1983) 'Mass transfer rate in ladle refining processes', *Scaninject III, Proc. 3rd Int. Conf. on Injection Metallurgy, Luleå, Sweden, 15–17 June 1983*, Mefos: Luleå, paper no. 12.)

Table 3.1, the value of n is, however, as high as 2 or 3. This can be attributed to the ejection of metal droplets at high gas flow rates which leads to an increase in $\mathcal{K}$ due to an increase in surface area. In some cases, the value of n is close to 0.25 for weakly agitated melts. This will be proven later (see Equation (3.42)).

3.2.2   *Models of interphase mass transfer*

For interpreting the results of a kinetic experiment, it is customary to assume that a model of mass transfer is operative at the interface.

Two-film model

The two-film model is the simplest model of mass transfer. As shown in Figure 3.2 the boundary films appear in each phase at the interface of two reacting phases. It is assumed that there is no hydrodynamic interaction between the two phases but that a condition of thermodynamic equilibrium exists at the interface. A steady-state mass transfer takes place in each boundary film such that, with reference to Figure 3.2, the flux of A in liquid, J_A, is given by

$$J_A = k_m^D(C_m^b - C_m^i) = k_g^D(C_g^i - C_g^b) \tag{3.34}$$

where k_m^D and k_g^D are the respective mass transfer coefficients in liquid and gas, C_m^b and C_g^b are the respective concentrations in the bulk liquid and gas, and C_m^i and C_g^i denote the corresponding concentrations at the interface. Since thermodynamic equilibrium exists at the interface, C_m^i and C_g^i are related to each other through the equilibrium constant or distribution ratio (also called the partition coefficient).

Boundary diffusion layer model

The boundary diffusion layer model was proposed essentially to explain the mechanism of mass transfer at the interface. It is assumed in this model that a thin diffusion boundary layer forms at the interface. The flow is split into two regions. In the diffusion layer adjacent to the interface, the turbulent diffusion is neglected and only molecular diffusion is assumed to take place. In the bulk, the contribution of

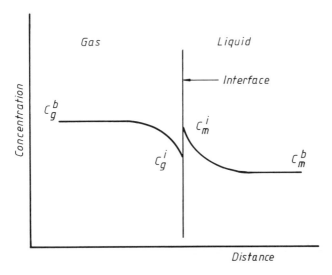

Figure 3.2 Concentration profiles at a gas–liquid interface.

turbulent diffusion is larger than the molecular diffusion. This situation may hold good, for example, in the case of a liquid flowing over a solid body and for which the effective thickness of the boundary layer, δ, is given by

$$\delta = D_m^{1/3} v^{1/6} \sqrt{(z/u)} \tag{3.35}$$

where v is the kinematic viscosity, u is the bulk liquid flow velocity and z is the distance from the point where liquid flows over the solid body.

It is impractical to employ the concept of effective film thickness in mass transfer calculations because k_m^D in Equation (3.9) is rarely proportional to D_m and δ varies as the diffusivity varies for different solutes.

Surface renewal model

The surface renewal model, also known as the Higbie model [2], is based upon the assumption of a moving film. Consider the interface of the slag and metal in Figure 3.3 in which a volume element of metal (phase 1) is carried by an eddy along the hypothetical path E_1 through E_4. At the positions E_2 and E_3 (at the interface) the element is assumed to be stagnant internally to allow unsteady-state diffusion or 'penetration' by the transferred species from phase 2, while at E_1 and E_4 (away from the interface) the elements are assumed to be completely mixed. Higbie showed that if the time spent by the element in the positions at the interface is t_i, then the diffusional mass transfer coefficient (k_m^D) within the element will be given by

$$k_m^D = 2\sqrt{(D_m/\pi t_i)} \tag{3.36}$$

Danckwerts (quoted in [2]) suggested that the constant time t_i that an element spends at the interface should be replaced by the average residence time evaluated from an assumed distribution pattern of particles at the surface. If t_i is expressed in

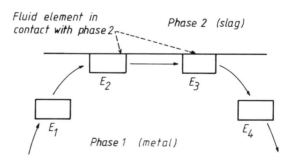

Figure 3.3   Schematic diagram of fluid motion in the Higbie model.

terms of a surface renewal factor S_f such that $t_i = 1/S_f$, then

$$k_m^D = \sqrt{(D_m S)} \tag{3.37}$$

The equations (3.36) and (3.37) are essentially of the same form and demonstrate that in an agitated melt the mass transfer coefficient is directly proportional to the square root of the molecular diffusion coefficient D_m. This has been found to hold good for a variety of experimental and industrial conditions.

For a melt covered by a viscous slag and weakly agitated by gas the value of A_D in Equation (3.11) will be nearly unity, and from Equation (3.36)

$$k_m^D = 2\sqrt{(D_m/\pi t_i)}$$

t_i is inversely related to the characteristic velocity:

$$t_i = L/u \tag{3.38}$$

where L is the characteristic length. Since D_m is constant

$$k_m^D \propto \sqrt{(u/L)} \tag{3.39}$$

On substituting the value of u from Equation (3.23), which corresponds to fluid motion dominated by the viscous force,

$$k_m^D \propto (\dot{\varepsilon}/\mu)^{1/4} \tag{3.40}$$

or

$$k_m^D \propto (Q/\mu)^{1/4} \tag{3.41}$$

(since $\dot{\varepsilon} \propto Q$). From Equation (3.33)

$$\mathcal{K} \propto \dot{\varepsilon}^{0.25} \tag{3.42}$$

which explains the coefficient of 0.25, as shown in Table 3.1.

3.2.3 *Mass transfer correlations*

The correlations for the mass transfer coefficient are often reported in terms of dimensionless numbers. Some important dimensionless numbers for mass transfer can be obtained as follows.

Let the left-hand side of Equation (3.5) be represented by

$$u_x \frac{\partial c}{\partial x} \simeq u_x \frac{\Delta c}{l} \tag{3.43}$$

and the right-hand side by

$$D_m \frac{\partial^2 c}{\partial x^2} \simeq D_m \frac{\Delta c}{l^2} \tag{3.44}$$

where l is the characteristic length. On equating Equations (3.43) and (3.44) one obtains a dimensionless number

$$Pe_m = \frac{u_x l}{D_m} = \frac{ul}{D_m} \tag{3.45}$$

which is called the Péclet diffusion number, Pe_m, for molecular transport (see also section 1.4.1). For eddy or turbulent transport, the Péclet turbulent number, Pe_t, is used:

$$Pe_t = \frac{u_x l}{D_t} = \frac{ul}{D_t} \tag{3.46}$$

By dividing the Péclet number by a dimensionless Reynolds number, ul/v, we obtain the Prandtl diffusion number, Pr_D, or Schmidt number, Sc:

$$\left(\frac{ul}{D_m}\right) \bigg/ \left(\frac{ul}{v}\right) = \frac{v}{D_m} = Sc \tag{3.47}$$

The Schmidt number expresses the ratio of physical constants v and D_m, i.e. the ratio of the rate of viscous momentum transfer to diffusive transfer. In physical terms, the velocity profile is determined by the kinematic viscosity while the concentration profile is determined by D_m. For gases, since D_m is high, the Schmidt number (v/D_m) is close to unity and thus the velocity and concentration profiles in a single-phase gas flow almost coincide with each other. In molten metals and slags the concentration profile is more elongated than the velocity profile (Figure 3.1) and therefore the Schmidt number is much greater than unity.

An important dimensionless number which relates the eddy diffusivity to the molecular diffusivity is the Nusselt diffusion number, Nu_D, or the Sherwood number, Sh, defined as

$$Sh = \frac{k_m l}{D_m} \simeq \frac{D_m + D_t}{D_m} \quad \left(\text{or } \frac{\text{total diffusivity}}{\text{molecular diffusivity}}\right) \tag{3.48}$$

The relationship between the Sherwood number, Reynolds number and Schmidt number, as shown in the following example, can be derived by using the method of indices.

Example 3.1 **Derive a relationship for the mass transfer coefficient in terms of the Sherwood, Reynolds and Schmidt dimensionless numbers by using the method of indices**

Suppose the mass transfer coefficient, k_m, is a function of fluid velocity (u), diffusion coefficient (D), kinematic viscosity (v) and the characteristic length (l) only:

$$k_m = Cu^a D^b v^c l^d \tag{A}$$

where C is the proportionality constant. On substituting the appropriate dimensions of length (L), mass (m) and time (t) into Equation (A):

$$L^1 t^{-1} = C(L^1 t^{-1})^a (L^2 t^{-1})^b (L^2 t^{-1})^c (L)^d$$

$$= C(L)^{a+2b+2c+d}(t)^{-a-b-c} \tag{B}$$

On equating the indices of L and t on the two sides for dimensional homogeneity:

$$a + 2b + 2c + d = 1$$

$$-a - b - c = -1$$

Eliminating a and b from the above two equations, we obtain $a = d + 1$ and $b = -(d + c)$. On substituting the values of a and b into Equation (A):

$$k_m = C(u)^{(d+1)}(D)^{-(d+c)}(v)^c(l)^d \tag{C}$$

On rearranging Equation (C) and substituting the values of the Schmidt and Reynolds numbers (from Equation (3.47)) and the Sherwood number (from Equation (3.48)):

$$\frac{k_m l}{D_m} = C\left(\frac{ul}{v}\right)^{d+1}\left(\frac{v}{D}\right)^{c+d+1}$$

$$Sh = C(Re)^{d+1}(Sc)^{c+d+1}$$

On substituting $m = d + 1$ and $n = c + d + 1$ we obtain

$$Sh = C Re^m Sc^n \tag{D}$$

Equation (D) is a typical correlation for mass transfer in terms of dimensionless numbers. The numerical values of the exponents m and n depend upon the system geometry and the fluid flow conditions.

Many correlations are reported in the literature for specific experimental and geometrical situations. Figure 3.4 shows, as an example, the pattern of dependence of the Sherwood number on flow conditions (i.e. Sh versus Re) within a single fluid phase flowing over a solid phase. In regime I, for low values of Re, the molecular viscous forces exceed the forces of inertia and $D_t = 0$. Thus, Sh is nearly constant. Regime I is called the molecular (laminar) transfer regime. Depending upon the relative magnitudes of D_m and D_t in Equation (3.48), with an increase in Re, Sh also increases. Regime II (intermediate regime) or regime III (turbulent exchange regime) is obtained at higher Re values. With a further increase in Re, both the turbulent viscosity and the diffusion coefficient attain their maximum values. The formation and the ejection of droplets from the liquid phase leads to an increase in mass transfer. Regime IV is, therefore, indicated by a dashed line. In order to take into account the influence of hydrodynamic factors, a hydrodynamic state factor can be introduced. For example, in a given geometrical situation the Sherwood number for one of the phases can be written as

$$Sh = Re^m Sc^n(1 + f) \tag{3.49}$$

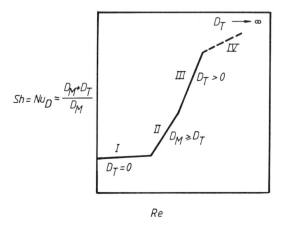

$$Sh = Nu_D \approx \frac{D_M + D_T}{D_M}$$

Figure 3.4   Typical variation of Sherwood number (*Sh*) with Reynolds number (*Re*).

where f accounts for the substance transported due to the interaction between phase flows. In addition to the hydrodynamic factors, the concentration, density or surface tension variations can lead to interfacial turbulence (Marangoni effect) which may lead to an unpredictable increase in the mass transfer rates. The effect of surface tension on mass transfer is discussed in section 3.4.

3.3   Chemical reaction

Consider a chemical reaction with stoichiometric equation $A + B \rightarrow AB$, which involves the collision or interaction of a single molecule of A with a single molecule of B to give a product molecule AB. The number of collisions of the molecules A and B is proportional to the rate of reaction, but the number of collisions at any given temperature is proportional to the concentration of reactants in the mixture. The rate of disappearance of A per unit area is given by

$$r_A = kC_A C_B \tag{3.50}$$

where k is the rate constant (also known as the specific rate) and C_A and C_B are the concentrations of A and B, respectively. The rate of a reaction and the rate constant depend on the way they are defined. For example, for the reaction $2A + B = 3C$, the molar rate of formation of C is three times the rate of disappearance of B:

$$\frac{dC_C}{dt} = -3\frac{dC_B}{dt} = -\frac{3}{2}\frac{dC_A}{dt} \tag{3.51}$$

The concepts of reversible and irreversible reactions, order of reaction, Arrhenius equation, absolute reaction rate theory and collision theory are briefly revised in Appendix A.

3.3.1   *Factors affecting the rate constant*

The chemical reaction rate constant (k) is strongly dependent upon temperature. According to the Arrhenius equation

$$k = \beta \exp(-E/RT) \tag{3.52}$$

where β is a proportionality constant and E is the activation energy for the chemical reaction. By evaluating k at different temperatures, one may plot $\lg k$ versus $\lg 1/T$ to determine E from the slope. For chemical reaction control, the value of E will be in the range of 2400–3600 kJ which is much higher than the activation energy (20–80 kJ) for a diffusion-controlled reaction. The mass transfer coefficient (k_m) does not vary much with temperature.

For some reactions, the chemical reaction rate constant (k) has been found to depend strongly upon the slag and the metal compositions. For example, the rate constant for sulphur transfer from slag to metal changes with the sulphide capacity of the slag (Figure 3.5). It has been reported that in the case of sulphur transfer there is a gradual transition from chemical reaction control for slags of low sulphide capacity to mass transport control in the metal phase for slags of high sulphide capacity. At some intermediate slag composition, k can become equal to k_m. In such cases it would be necessary to find k at different temperatures and determine the activation energy to differentiate between mass transfer and chemical reaction as the rate-limiting steps. In the case of chemical reaction control, the fluid flow or stirring conditions do not affect the rate. The presence of some alloying elements may affect the rate-controlling mechanism. For example, it has been suggested [3] that in the case of sulphur transfer from molten iron to a 49% CaO–45% SiO$_2$–6% MgO slag, the rate constant for chemical reaction, k, was smaller than the mass transfer

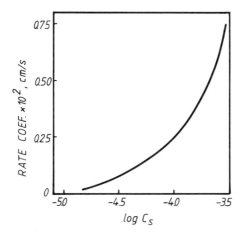

Figure 3.5   Variation of the rate coefficient of sulphur transfer with sulphide capacity of slag. (*Source*: Deo, B. and Grieveson, P. (1986) *Steel Res.*, **57**, 514–19.)

coefficient. The value of k was found to increase with rising silicon concentration in the metal. At silicon contents greater than 0.8 wt %, the rate of desulphurization was controlled by mass transport in the metal phase only.

3.4 Adsorption and surface tension

Adsorption is one of the five possible rate-limiting steps in the overall reaction (see section 3.1). Adsorption is of two kinds:

1. Physical adsorption The force responsible for physical adsorption is purely physical (Van der Waals) attraction. The heat of adsorption does not exceed a few kilojoules per mole of the adsorbed species. It is thus weak in nature and dominant only at low temperatures.

2. Chemisorption Specific chemical forces exist between the adsorbed species and the substrate. The heat evolved as a result of adsorption is of the order of 40 to 400 kJ/mol. Chemisorption is irreversible; for example, oxygen chemisorbed on charcoal yields carbon monoxide gas. Some special features of chemisorption are:

    (a) the surface becomes saturated after it is covered by a layer (10–100 nanometres thick) of adsorbed molecules, so further adsorption takes place with difficulty;

    (b) chemisorption is associated with an appreciable activation encrgy and therefore it is a relatively slow process, which renders it difficult at low temperatures;

    (c) since solid surfaces are not smooth and uniform, some surface sites are chemically more active than the others;

    (d) there are interactions, usually of a repulsive nature, between the atoms or molecules adsorbed side by side on a surface.

The relationship between the equilibrium amount of adsorption and the pressure (or concentration) at constant temperature is called an adsorption isotherm. Different types of isotherms are described in the literature but the Langmuir isotherm is frequently used for liquid iron alloys.

3.4.1 *Langmuir isotherm*

The Langmuir isotherm is an ideal isotherm which applies to chemisorption on a perfectly uniform surface without any interaction amongst the adsorbed molecules. Suppose the gas (G) is adsorbed undissociated on a base surface site (S); the adsorption and desorption processes can then be represented as

$$G + S \Leftrightarrow \overset{\displaystyle G}{\underset{\displaystyle S}{|}} \qquad\qquad (3.53)$$

Let θ be the fraction of covered surface sites and thus $(1 - \theta)$ the fraction of the base or uncovered surface sites. Then the rate of adsorption is $k_f p(1 - \theta)$ and the rate of desorption is $k_b \theta$, where k_f, k_b are the forward and backward rate constants and p is the pressure of gas G.

At equilibrium

$$k_f p(1 - \theta) = k_b \theta$$

or

$$\frac{\theta}{1 - \theta} = \frac{k_f}{k_b} p = K_e p \tag{3.54}$$

where K_e is the adsorption equilibrium constant. The value of θ can be calculated as

$$\theta = \frac{K_e p}{1 + K_e p} \tag{3.55}$$

At low pressure

$$\theta \simeq K_e p \tag{3.56}$$

At high pressure $\theta \simeq 1$ or

$$(1 - \theta) \simeq \frac{1}{K_e p} \tag{3.57}$$

Note that in this treatment a bare site is treated as a reactant.

3.4.2 Adsorption with dissociation

Adsorption is preceded sometimes by dissociation of a molecule on the surface. For example, hydrogen is adsorbed on metals in the atomic form. It can be shown that if the adsorbed molecule dissociates into two atoms, then

$$(1 - \theta) = \frac{1}{1 + K_e^{1/2} p^{1/2}} \tag{3.58}$$

If $K_e^{1/2} p^{1/2} \gg 1$, the fraction of surface that is bare is inversely proportional to the square root of pressure.

3.4.3 Effects of adsorption on surface tension and reaction kinetics

Preferential adsorption at the surface can lead to a change in the surface tension. The dissolved elements in molten iron, like oxygen, selenium and sulphur, are known to lower the surface tension considerably. Figure

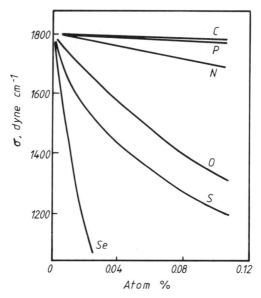

Figure 3.6   Surface tension of liquid iron as a function of the concentration (at.%) of dissolved solutes carbon, phosphorus, nitrogen, oxygen, sulphur and selenium. (*Source*: Kozakevitch, P. (1968) *Surface phenomena of metals*, Society of Chemical Industry Monograph 28, p. 223.)

3.6 shows the surface tension for binary solutions of carbon, phosphorus, oxygen, nitrogen, sulphur and selenium in liquid iron at 1823 K. In a multicomponent solution, however, the net value of surface tension can change unexpectedly with the concentration of solutes. The total effect, unlike in the evaluation of the activity coefficient in a multicomponent solution, is not additive; for example, carbon has virtually no surface activity but it raises the activity coefficient of dissolved sulphur. With an increase in the carbon content the surface tension of iron will decrease. Chromium and carbon are both non-active solutes, but at low concentrations they form free surfactive carbides, thereby lowering the surface tension. This phenomenon does not take place when tungsten and carbon, which also form carbides, are present in solution in molten iron. There are also no dramatic changes in the surface tension when chromium and nitrogen, or vanadium and nitrogen, are present together, despite the fact that they form stable nitrides.

The excess of solute in the surface region, Γ_s, over and above the concentration in the bulk liquid can be calculated from Gibbs' adsorption equation:

$$\Gamma_s = -\frac{1}{RT}\left(\frac{\partial \sigma}{\partial (\ln a_s)}\right)_T \tag{3.59}$$

where σ is the surface tension and a_s is the activity of the solute in the bulk phase. The plot of Γ_s for oxygen and sulphur [4] is shown in Figure 3.7. Beyond about

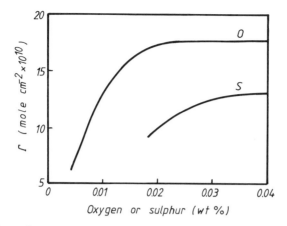

Figure 3.7 Effect of oxygen and sulphur concentration on the excess of solute in the surface region. (*Source*: Kozakevitch, P. (1968) *Surface phenomena of metals*, Society of Chemical Industry Monograph 28, p. 223.)

0.03 wt % sulphur or 0.02 wt % oxygen, Γ_s becomes constant. If this is assumed to correspond to a complete coverage of the surface by the adsorbed species then, at lower concentrations, we can show that the fraction of covered surface is proportional to the bulk concentration. Similar to Equation (3.54), for oxygen

$$[O]_b + o_s = [O]_s$$

$$K_e = \frac{[O_s]}{[O]_b o_s} = \frac{\theta}{(1 - \theta)[O]_b} \tag{3.60}$$

where o_s is a vacancy on the surface, subscript b refers to bulk and subscript s refers to surface. At 1823 K, the value of K_e appearing in Equation (3.60) equals 780 at saturation [4]. Assuming that oxygen is adsorbed as an ideal monolayer on the surface of liquid iron, the dashed line in Figure 3.8 shows that the activity coefficient of oxygen in the surface layer decreases markedly as its concentration increases. In other words, more and more oxygen atoms try to reach the interface. The solid line in Figure 3.8 corresponds to the value of $K_e = 71$ at 1823 K [4].

An increase in temperature lowers the surface tension. The efficacy of surfactive solutes in lowering surface tension thus decreases with a rise in the temperature of molten iron. The heat of adsorption (ΔH^{adsorb}) can be derived from

$$\left(\frac{\partial(\ln a)}{\partial(1/T)}\right)_T = -\left(\frac{\Delta H^{adsorb}}{R}\right) \tag{3.61}$$

The kinetics of adsorption has so far been difficult to predict. It is faster than predicted by diffusion in the metal phase alone. For example, in the case of oxygen adsorbed on the surface of molten iron, the equilibrium concentration is established instantaneously, even when the bulk concentration is one-tenth of its saturation

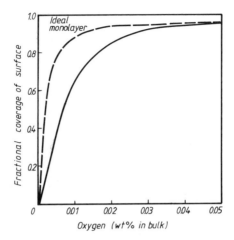

Figure 3.8 Fractional coverage of surface as a function of oxygen concentration (wt%) in bulk. (*Source*: Swisher, J. H. and Turkdogan, E. T. (1967) *Trans. Metall. Soc. AIME*, **239**, 602–10.)

concentration. In one extreme, the surfactive solute can affect the kinetics of gas–metal reactions by retarding the circulation of liquid around a rising gas bubble, thereby reducing the rate of mass transfer. Surfactive solutes can also prevent the eddies from renewing the surface. At the other extreme, they can cause interfacial turbulence (Marangoni effect) and enhance mass transfer. Such effects, occurring in slag–metal–gas systems at high temperature, are not yet amenable to kinetic treatment.

3.5 Kinetics of slag–metal reactions in a gas-stirred bath

The slag–metal reactions in steelmaking can be stimulated by gas stirring. Consider the simple case of the transfer of sulphur from metal to slag. Figure 3.9 shows some typical concentration gradients in a slag–metal system. The interfacial and bulk concentrations are indicated by superscripts i and b, respectively. Five separate cases are as follows:

Case 1 (Figure 3.9a): Mass transfer of sulphur in the metal phase is rate controlling. Equilibrium exists at the interface.

Case 2 (Figure 3.9b): Mass transfer of sulphur in the slag phase is rate controlling. Equilibrium exists at the interface.

Case 3 (Figure 3.9d): Chemical reaction at the interface is rate controlling. Equilibrium does not exist at the interface.

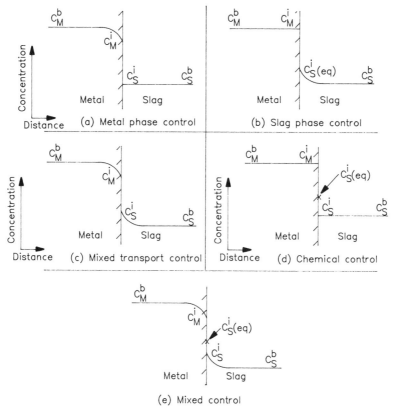

Figure 3.9 The five possible rate-controlling mechanisms in the slag–metal reaction: metal phase control (a); slag phase control (b); mixed mass transport control (c); chemical control (d); and (e) mixed control. (*Source*: Richardson, F. D. (1974) *Physical Chemistry of Melts in Metallurgy*, Vol. 2, Academic Press: London, p. 420.)

Case 4 (Figure 3.9c): Mixed control (chemical reaction plus mass transfer) in metal and slag. Equilibrium does not exist at the interface.

Case 5 (Figure 3.9e): Mixed mass transfer control in metal and slag. Equilibrium exists at the interface.

In the following discussion it is assumed that the interfacial tension does not change significantly while the mass transfer or chemical reaction is in progress.

Case 1: Metal phase control

In a gas-stirred system the equation for molar flux J_D, i.e. number of moles transferred per unit time per unit area, can be written from Equation (3.7) as

$$J_D = \text{diffusive flux} \ (J_{D_m}) + \text{momentum flux} \ (J_{D_t})$$

The momentum flux approaches a zero value at the interface, and from Equation (3.10)

$$J_D = J_{D_m} = k_m^D(C_m^b - C_m^i)$$

where $k_m^D = k_m$ from Equation (3.11) when $A_D = 1$.

At the temperatures of steelmaking, the chemical reactions are extremely fast. Therefore, a pseudo or instantaneous thermodynamic equilibrium may be assumed to exist at the interface during the small time interval ($\Delta t \to 0$), such that (with reference to Figure 3.9a):

$$\frac{C_m^{eq}}{C_s^{eq}} = \frac{C_m^i}{C_s^i} = L_p \tag{3.62}$$

where L_p is the equilibrium constant, sometimes denoted as the partition or distribution coefficient; C_m^{eq} and C_s^{eq} are the ultimate thermodynamic equilibrium concentrations in metal and slag, respectively.

Substituting Equation (3.62) into Equation (3.10):

$$J_D = J_m = k_m(C_m^b - L_p C_s^i)$$

From Figure 3.9(a), for metal phase control, $C_s^i = C_s^b$ and hence

$$J_m = -\frac{dC}{dt}\frac{V_m}{A} = k_m(C_m^b - L_p C_s^i) \tag{3.63}$$

where V_m is the volume of the metal (m³) and A is the nominal area of the interface (m²). The use of Equation (3.63) to evaluate the mass transfer coefficient will be evident from the following example.

Example 3.2 Evaluate the mass transfer coefficient assuming metal phase control during sulphur transfer from metal to slag

Suppose $CaO-Al_2O_3-CaF_2$ flux is laid on the surface of a completely stirred molten iron–sulphur alloy at 1583 K. The metal weighs 4 kg and contains 0.1 wt % sulphur at time $t = 0$; the inner diameter of the crucible is 0.07 m. Metal samples are taken at regular time intervals and analyzed for sulphur:

Time (s)	Wt % sulphur in iron
0	0.1
60	0.08
120	0.06
240	0.04
480	0.02

The density of liquid iron is 7000 kg/m³. The equilibrium concentration of sulphur in the metal is 0.005 wt %. Assuming the mass transfer of sulphur in the metal phase to be rate limiting, it is required to determine the mass transfer coefficient.

Let

$S_0 = $ wt % sulphur at time 0

$S_t = $ wt % sulphur at time t

$S_e = $ wt % sulphur at equilibrium

$\rho_m = $ density of metal

$\rho_s = $ density of slag

Let the weight of slag be WS and the weight of metal be WM. If the slag does not contain any sulphur to start with, then from mass balance the molar concentrations C_s^b, C_m^b, C_s^{eq}, C_m^{eq} appearing in Equations (3.62) and (3.63) can be evaluated as

$$C_s^b = \frac{WM(S_0 - S_t)}{WS}\left(\frac{\rho_s}{100}\right)\left(\frac{1}{32}\right)$$

$$C_m^b = \rho_m\left(\frac{S_t}{32}\right)\left(\frac{1}{100}\right)$$

$$C_m^{eq} = \rho_m\left(\frac{S_e}{32}\right)\left(\frac{1}{100}\right)$$

$$C_s^{eq} = \frac{WM(S_0 - S_e)}{WS}\left(\frac{\rho_s}{100}\right)\left(\frac{1}{32}\right)$$

$$\frac{C_m^{eq}}{C_s^{eq}} = L_p$$

Substituting these into Equation (3.63) and integrating:

$$-\ln\left(\frac{S_t - S_e}{S_0 - S_e}\right) = k_m A \frac{\rho_m}{WM} t \left(\frac{S_0}{S_0 - S_e}\right) \tag{A}$$

From the experimental data given above

$S_e = 0.005$ wt %

$S_0 = 0.1$ wt %

$A = \pi(0.035)^2$ m^2

$\rho_m = 7000$ kg/m^3

$WM = 4$ kg

Now, from the slope of the straight line in the graphical plot of

$-\ln[(S_t - S_e)/(S_0 - S_e)]$

versus time, the value of k_m can be estimated as 5.3×10^{-4} m/s. This is also the order of magnitude of mass transfer coefficients in gas-stirred slag–metal systems.

Case 2: Slag phase control

The procedure for deriving the rate equation for slag phase control is exactly the same as described in case 1. The flux in the slag phase (J_s) can be written as

$$J_s = k_s(C_s^i - C_s^b) \tag{3.64}$$

where k_s is the mass transfer coefficient in the slag phase. At equilibrium

$$\frac{C_m^{eq}}{C_s^{eq}} = \frac{C_m^i}{C_s^i} = \frac{C_m^b}{C_s^i} = L_p \tag{3.65}$$

Substituting Equation (3.65) into Equation (3.64):

$$J_s = -k_s\left(C_s^b - \frac{C_m^b}{L_p}\right) \tag{3.66}$$

The concentrations C_s^b and C_m^b can be determined by analyzing the metal samples taken at regular intervals. For the known values of L_p and k_s, J_s can be found at each instant.

The conditions under which the predominance of slag phase control may arise are as follows:

1. The bulk slag phase is presaturated with the element to be transferred from the metal and the concentration of the element in the metal phase is high. Such a situation can also be artificially created when the layer of the reaction product formed on the slag side is such that it does not allow proper homogenization of the bulk slag phase, even in the presence of strong stirring conditions. For example, consider the desulphurization of hot metal by injection of a powdered lime-based flux. It has been experimentally observed that a layer of solid CaS forms at the lime–metal interface, thereby hindering the further transfer of sulphur from the metal to the slag. Another example is the reaction

$$2\langle CaO\rangle + 2[S] + [Si] = 2\langle CaS\rangle + \langle SiO_2\rangle \tag{3.67}$$

   In addition to CaS, a silica-enriched solid layer may form at the interface hindering the transport of sulphur in the slag.
2. The physical conditions of stirring in the viscous (or semi-solid) slag phase are such that $k_s \ll k_m$ and, in addition to this, the slag has low sulphide capacity (i.e. a high value of L_p).
3. The slag acts as a medium or source of the element to be transferred to the metal, for example the transfer of oxygen from slag of low oxygen potential (i.e. containing very small amounts of FeO) to the liquid iron having virtually no dissolved oxygen.

Case 3: Chemical reaction control

With reference to Figure 3.9(d):

$$\text{rate of forward reaction} = k_f C_m^i = k_f C_m^b \tag{3.68}$$

Since the concentration gradient between the bulk and the interface disappears when the chemical reaction is rate controlling, $C_m^i = C_m^b$ and also $C_s^i = C_s^b$. Now

$$\text{rate of backward reaction} = k_b C_s^i = k_b C_s^b \tag{3.69}$$

Hence the net rate is

$$-\frac{dC}{dt}\frac{V_m}{A} = (k_f C_m^b - k_b C_s^b) \tag{3.70}$$

At equilibrium, if the respective concentrations in metal and slag are C_m^{eq} and C_s^{eq} and the net rate is zero,

$$k_f C_m^{eq} = k_b C_s^{eq} \tag{3.71}$$

If the equilibrium constant or the partition coefficient is L_p, similar to Equation (3.62),

$$\frac{C_m^{eq}}{C_s^{eq}} = \frac{k_b}{k_f} = L_p \tag{3.72}$$

Substituting Equation (3.72) into Equation (3.70):

$$-\frac{dC}{dt}\frac{V_m}{A} = k_f(C_m^b - C_s^b L_p) \tag{3.73}$$

It can be seen that Equation (3.73) is identical in form to Equation (3.63) except that k_f is the chemical reaction rate constant and k_m is the mass transfer coefficient. It is possible only through experiment to ascertain whether a reaction is controlled by mass transfer, or by chemical reaction, or by both at the interface.

Case 4: Mixed control

Most of the liquid–liquid reactions in steelmaking are controlled by mass transfer. The conditions of mixed control, i.e. mass transfer plus chemical reaction control, arise in gas–metal and slag–metal–gas reactions and will be discussed later (see, for example, the nitrogen adsorption reaction in section 3.6.1 and the CO_2–C–O reaction in Chapter 6).

Case 5: Mixed mass transfer control in metal and slag

The slag–metal reactions can be viewed as coupled reactions without specifying a priori (or even verifying) either slag phase control or metal phase control as the rate-limiting steps. On equating Equations (3.63) and (3.64), the interfacial concentration (C_s^i) can be eliminated to show that

$$J_s = \frac{L_p C_s^b - C_m^b}{L_p/k_s + 1/k_m} \tag{3.74}$$

The denominator can be regarded as the total resistance to mass transfer where L_p/k_s is the resistance in the slag phase and $1/k_m$ is the resistance in the metal phase.

For large values of L_p the slag phase resistance L_p/k_s becomes much greater than $1/k_m$. Generally, $k_m/k_s < 10$, and for small values of L_p the slag–metal reactions in steelmaking are controlled by mass transfer in the metal phase alone. However, depending upon the process conditions and the instantaneous interfacial values of L_p/k_s and $1/k_m$, the process can switch over from metal phase control to mixed transport control or vice versa.

The multiplicity of reactions and the corresponding number of mass transfer coefficients increase with the number of reacting components and the phases involved. To tackle this situation a completely generalized multicomponent mixed transport control theory has been developed involving both oxide- and sulphide-forming reactions. This theory has been successfully applied to the modelling of the desulphurization of hot metal [5], the simultaneous desulphurization and dephosphorization of hot metal [6], and to the dynamic control of the oxygen steelmaking process [7]. The treatment given below is adapted from Robertson *et al.* [5].

3.5.1 *Coupled transfers of oxygen, silicon, manganese, iron, sulphur and phosphorus between slag and metal*

The reactions occurring at the slag–metal interface are coupled reactions of the type

$$[M] + \sigma[O] = M^{2\sigma+} + \sigma O^{2-} = MO_\sigma \tag{3.75}$$

and

$$[M] + \sigma[S] = M^{2\sigma+} + \sigma S^{2-} = MS_\sigma \tag{3.76}$$

where σ is a stoichiometric constant.

The metallic [M] components can be considered to include iron, phosphorus, manganese, silicon, aluminium and calcium. A feature of the theory which illustrates its general applicability is that none of the oxides or sulphides are considered as inert. The extent to which each component takes part in the reactions is determined in the light of the relevant equilibria and the flux density equations. The chemical kinetics of all these reactions is assumed to be fast.

Sulphur is considered to transfer from the bulk of the metal to the interface and then into the slag as sulphide ions. Oxide ions transfer counter to sulphide ions and positive $M^{2\sigma+}$ ions may transfer in either direction. The actual driving forces used are the molar concentration differences for the oxides and sulphides between the interface and the bulk. The bulk slag and metal are assumed to be thoroughly mixed such that no concentration gradients exist in the bulk. Electroneutrality for oxide and sulphide ion transfer is assured by associating these ions with the $M^{2\sigma+}$ ions according to the stoichiometry and to the chemical potentials of the various oxides and sulphides. This procedure still implies the ionic nature of the slag phase. The effect of an applied cathodic potential and the corresponding electric current density across the slag–metal interface can, in principle, be included in the theory. This can be done through their effect on the equilibrium constants and flux density equations, respectively, provided that the anode reactions are also understood.

Equilibria

For each reaction at the slag–metal interface, it is possible to write the relevant equilibrium constants (K) which relate the interfacial concentrations of the components (subscript i for interface) because the reactions are assumed to be 'fast'. For each component

$$K_1 = \frac{(a_{MO_\sigma})}{a_M [a_O]^\sigma} \tag{3.77}$$

$$K_2 = \frac{(a_{MS_\sigma})}{a_M [a_S]^\sigma} \tag{3.78}$$

where MO and MS denote the oxide and sulphide, respectively, formed by the element M; a refers to the Raoultian activity and S and O refer to sulphur and oxygen, respectively. Slag and metal interface concentrations can thus be related by the equations

$$OX_i = KOX(X_i O_i^\sigma) \tag{3.79}$$

$$SU_i = KSU(X_i S_i^\sigma) \tag{3.80}$$

where OX_i and SU_i represent, respectively, the oxide and sulphide mole fractions of the component in the slag at the interface; similarly, X_i, O_i and S_i represent the relevant mole fractions in the metal at the interface.

For any component

$$KOX = \left(\frac{\gamma_M \gamma_O^\sigma}{\gamma_{OX}} \right) K_1 \tag{3.81}$$

$$KSU = \left(\frac{\gamma_M \gamma_S^\sigma}{\gamma_{SU}} \right) K_2 \tag{3.82}$$

The γ_{OX} and γ_{SU} values are Raoultian activity coefficients of oxides and sulphides, respectively, in the slag, while the γ_M, γ_S and γ_O are the activity coefficient values for dilute components in the metal. Some of the equilibrium constant values are given in Table 3.2.

Table 3.2   Stoichiometric constants σ and
thermodynamic equilibrium constant K at 1738 K.

	Components		
	Fe	Si	Ca
σ	1	2	1
K_1 (oxide)	553.4	5.15×10^{13}	7.32×10^{14}
K_2 (sulphide)	1.75	3.53	1.14×10^{11}

Flux density equations

There are n flux density equations for the transfer of the n metallic components and two more for oxygen and sulphur. For the metallic components:

$$\text{transfer rate out of metal} = \text{transfer rate into slag} \tag{3.83}$$

i.e.

$$k_{mM} C_{Vm}(X_b - X_i) = k_{OX} C_{Vs}(OX_i - OX_b) + k_{SU} C_{Vs}(SU_i - SU_b) \tag{3.84}$$

where C_{Vm} and C_{Vs} are the molar volumes of metal and slag, respectively, k_{mM} is the mass transfer coefficient of the elements in the metal phase (assumed to be the same for all elements, for the sake of simplicity), k_{OX} and k_{SU} are the mass transfer coefficients of the oxides and the sulphides, respectively, in the slag phase, and, as before, b refers to bulk and i to the interface.

Equation (3.84) can be rearranged to give

$$X_i = TM(k_{OX}^* KOXO_i^\sigma + k_{SU}^* KSUS_i^\sigma + 1) \tag{3.85}$$

where TM is a 'total concentration' of any component defined by

$$TM = (X_b + k_{OX}^* OX_b + k_{SU}^* SU_b) \tag{3.86}$$

The k^* values are defined for each component as

$$k_{OX}^* = \frac{k_{OX} C_{Vs}}{k_{mM} C_{Vm}} \tag{3.87}$$

$$k_{SU}^* = \frac{k_{SU} C_{Vs}}{k_{mM} C_{Vm}} \tag{3.88}$$

Equation (3.85) is important because it shows that it is possible to express the surface concentration (activity) of any component in terms of the bulk concentration in the metal and slag phases, including the O and S concentrations.

The oxygen flux density equation, on adding the contribution from all the n oxides in the slag, is

$$C_{Vm} k_{mO}(O_b - O_i) = \sum_{i=1}^{n} \sigma C_{Vs} k_{OX}(OX_i - OX_b) \tag{3.89}$$

where k_{mO} is the mass transfer coefficient of oxygen in metal. The values of OX_i can be written in terms of X_i and O_i and the X_i in turn can be written in terms of O_i and S_i by using Equation (3.85). An equation with only two unknowns, O_i and S_i, is thus obtained.

Another equation with these two unknowns can be obtained from the sulphur flux density equation which is directly analogous to Equation (3.89):

$$C_{Vm}k_{mS}(S_b - S_i) = \sum_{i=1}^{n} \sigma C_{Vs}k_{SU}(SU_i - SU_b) \tag{3.90}$$

where k_{mS} is the mass transfer coefficient of sulphur in metal. Thus, a set of two simultaneous equations with two unknowns are obtained. If sulphur transfer is not considered, only one equation in one unknown, O_i, will be obtained. These equations can be solved numerically for the known initial values of the bulk concentrations to give O_i and S_i. Once these values are obtained, the X_i values can be calculated using Equation (3.85). The flux density of each component is then calculated as

$$J_X = k_{mM}(X_b - X_i)C_{Vm} \tag{3.91}$$

$$J_O = k_{mO}(O_b - O_i)C_{Vm} \tag{3.92}$$

$$J_S = k_{mS}(S_b - S_i)C_{Vm} \tag{3.93}$$

Conservation equations

It is now possible to calculate a new bulk metal composition at some later time. The change in bulk composition (X_b) over a time interval Δt is given by the equation: rate of removal = flux. If V_m is the volume of metal

$$C_{Vm}V_m\Delta(X_b)/\Delta t = -Ak_{mM}C_{Vm}(X_b - X_i) \tag{3.94}$$

Using a dimensionless time defined as

$$t^* = (A/V_m)tk_{mS} \tag{3.95}$$

$$\Delta(X_b) = \Delta t^*(X_b - X_i)(k_{mM}/k_{mS}) \tag{3.96}$$

the change can be expressed as

$$X_{b(new)} = X_{b(old)} + \Delta(X_b) \tag{3.97}$$

The progress of the reactions can be followed by taking small time steps (say $\Delta t = 0.01$ seconds). By comparing the predicted rates in t with actual rates in real time, the value of (Ak_{mM}/V_m) can be estimated from which knowledge of A/V_m allows k_{mM} to be estimated.

Changes in oxide concentrations can be estimated using the flux density equations for each component:

$$J_{OX} = k_{OX}C_{Vs}(OX_i - OX_b) \tag{3.98}$$

and the conservation equations. If V_s is the volume of slag

$$C_{Vs} V_s \frac{\Delta(OX_b)}{\Delta t} = k_{OX} A C_{Vs} (OX_i - OX_s) \tag{3.99}$$

Using the same definition of t^* as in Equation (3.95):

$$\Delta(OX_b) = \Delta t^* \left(\frac{k_{OX}}{k_{mS}} \right) \left(\frac{V_m}{V_s} \right) (OX_i - OX_b) \tag{3.100}$$

An equation analogous to Equation (3.100) can be written for the sulphides.

Generally, slag analyses are carried out for components M in the slag and not for sulphides and oxides separately. The change $\Delta(X_s)$ in the slag can be simply calculated from the change $\Delta(X_b)$ in the metal, using the equation

$$-\Delta(X_s) = \left(\frac{C_{Vm}}{C_{Vs}} \right) \left(\frac{V_m}{V_s} \right) \Delta(X_b) \tag{3.101}$$

It can be assumed, for the sake of simplicity, that for all the components in the slag, $k_{SU} = k_{OX} = k_s$. In particular systems, where there is sufficient information on diffusivities, it may be justifiable to choose particular k values for each component in both phases.

The equations given above allow the calculation, as a function of dimensionless time t^*, of the change in composition of the slag and metal phases as the multicomponent transport processes occur by means of the coupled reactions.

As an example of the application of mixed transport control theory, the kinetic model of reactions in the $(Fe-Mn-O)-(FeO-MnO)$ slag–metal system is developed in the example given below.

Example 3.3 **Application of mixed mass transport control theory to the kinetics of oxidation/deoxidation reactions in the $(Fe-Mn-O)-(FeO-MnO)$ system**

For the sake of simplicity, let us assume that FeO and MnO form an ideal solution in slag, and that manganese and oxygen, dissolved in iron, obey Henry's law.

The manganese–oxygen reaction occurring at the slag–metal interface is $[Mn] + [O] = (MnO)$. The equilibrium constant can be written as

$$K_{MnO} = \frac{a^i_{MnO}}{[h^i_{Mn}][h^i_O]} = \frac{[X^i_{MnO}]}{[X^i_{Mn}][X^i_O]} \tag{A}$$

where, as before, superscript i refers to the interfacial concentration. Similarly, the iron–oxygen reaction at the slag–metal interface is $[Fe] + [O] = (FeO)$. The equilibrium constant can be written as

$$K_{FeO} = \frac{a^i_{FeO}}{[a_{Fe}][h^i_O]} = \frac{X^i_{FeO}}{[X^i_{Fe}][X^i_O]} \tag{B}$$

The flux density and conservation equations can be written as follows.

In the case of manganese, the transfer rate of manganese out of the metal equals the transfer rate of MnO into the slag, $J_{Mn} = J_{MnO}$, or

$$C_{Vm}k_{Mn}(X_m^b - X_{Mn}^i) = C_{Vs}k_{MnO}(X_{MnO}^i - X_{MnO}^b) \quad \text{(C)}$$

The value of X_{MnO}^i can be substituted from Equation (A):

$$C_{Vm}k_{Mn}(X_{Mn}^b - X_{Mn}^i) = C_{Vs}k_{MnO}(K_{MnO}X_{Mn}^i X_O^i - X_{MnO}^b) \quad \text{(D)}$$

Similarly, from Equation (D), we can write for iron

$$C_{Vm}k_{Fe}(X_{Fe}^b - X_{Fe}^i) = C_{Vs}k_{FeO}(K_{FeO}X_{Fe}^i X_O^i - X_{FeO}^b) \quad \text{(E)}$$

The flux of oxygen is related to the flux of iron and manganese:

$$C_{Vm}k_{Mn}(X_{Mn}^b - X_{Mn}^i) + C_{Vm}k_{Fe}(X_{Fe}^b - X_{Fe}^i) = C_{Vm}k_O(X_O^b - X_O^i) \quad \text{(F)}$$

The equations (D)–(F) constitute three non-linear equations in the three unknowns X_O^i, X_{Fe}^i and X_{Mn}^i (provided k_O, k_{Mn}, k_{Fe}, k_{FeO} and k_{MnO} are known). These non-linear equations can be solved by suitable numerical methods. We then can proceed with the computation of new bulk concentrations by evaluating J_{Mn}, J_{Fe} and J_O at each time step:

$$J_{Mn} = C_{Vm}k_{Mn}(X_{Mn}^b - X_{Mn}^i) \quad \text{(G)}$$

or

$$\frac{C_{Vm}V_m \Delta M_{Mn}}{\Delta t} = -A k_{Mn} C_{Vm}(X_{Mn}^b - X_{Mn}^i) \quad \text{(H)}$$

where V_m is the volume of metal and ΔM_{Mn} are the number of moles of manganese transferred from the metal to the slag phase. On simplification

$$\Delta M_{Mn} = -\frac{A}{V_m}(\Delta t)k_{Mn}C_{Vm}(X_{Mn}^b - X_{Mn}^i) \quad \text{(I)}$$

$$|\Delta M_{Mn}| = |\Delta M_{MnO}| \quad \text{(for the slag phase)} \quad \text{(J)}$$

Similarly, in the case of iron,

$$\Delta M_{Fe} = -\frac{A}{V_m}(\Delta t)k_{Fe}C_{Vm}(X_{Fe}^b - X_{Fe}^i) \quad \text{(K)}$$

$$|\Delta M_{Fe}| = |\Delta M_{FeO}| \quad \text{(L)}$$

and for oxygen

$$\Delta M_O = -\frac{A}{V_m}(\Delta t)k_O C_{Vm}(X_O^b - X_O^i) \quad \text{(M)}$$

The values of ΔM_O, ΔM_{Fe}, ΔM_{Mn}, ΔM_{MnO}, ΔM_{FeO} can be used to calculate the new values of bulk metal and slag compositions at each time step, Δt. We can repeat the iterative calculation procedure to calculate new bulk concentrations for the next time step Δt and so on, till equilibrium is achieved.

The significance of the term A/V_m should be noted. It signifies that if metal is present as a droplet in the slag phase, or vice versa, then the total effective surface area will be multiplied by the number of droplets present and the kinetics of the reaction would proceed towards equilibrium very rapidly. This may, for example, occur when metal droplets are ejected into the slag phase by an impinging oxygen jet in the oxygen steelmaking converter (see Chapters 5 and 6).

3.5.2 *Mass transfer correlations for gas-stirred slag–metal systems*

Some of the empirical correlations between the capacity coefficient of mass transfer, $\mathcal{K}$ (as defined in Equation (3.32)), and mixing power density for gas-stirred liquid–liquid systems are listed in Table 3.1. Cold model studies [8] on aqueous solution–amalgam systems as well as molten salt–lead systems for moderate gas flow rates (i.e. when metal does not penetrate the slag–metal interface due to turbulence or the ejection of metal droplets in the slag) led to the following empirical correlation for the metal side mass transfer coefficient k_m:

$$k_m = \alpha \sqrt{(DQ/A)} \tag{3.102}$$

where D is the diffusion coefficient of the species, A is the nominal slag–metal interface area and Q is the volumetric flow rate of gas. The value of α for laboratory-scale experiments is $1\,\mathrm{m}^{-1/2}$; owing to geometrical scaling and turbulence in the system, the value of α for the oxygen steelmaking converters is approximately $5\,\mathrm{m}^{-1/2}$.

Several correlations are also reported in terms of the dimensionless numbers. The apparent mass transfer coefficient between a mercury layer and an aqueous layer agitated by bubbles is given by [1]

$$Sh = 1.33 Re^{0.7} Sc^{1/3} \tag{3.103}$$

If Equations (3.23) and (3.25) are substituted into Equation (3.103), then the relationship between the apparent mass transfer coefficient, k_m, and $\dot{\varepsilon}$ is obtained, respectively, as

$$k_m \propto Sc^{-2/3} \rho^{-0.533} L^{0.07} \dot{\varepsilon}^{0.233} \tag{3.104}$$

and

$$k_m \propto Sc^{-2/3} \rho^{-0.533} L^{0.4} l^{-0.47} \dot{\varepsilon}^{0.233} \tag{3.105}$$

This reconfirms the exponent of 0.25 for $\dot{\varepsilon}$ (see Table 3.1).

The dimensionless numbers frequently appear in mass transfer correlations. With the help of these correlations the results obtained on the laboratory scale can be extended to the industrial scale. Let us examine, as an example, the mass transfer correlations proposed [9, 10] for the oxidation of silicon dissolved in metal by a slag containing FeO, on the basis of laboratory-scale experiments involving a gas-stirred slag–metal system at elevated temperatures (1523 K). It will serve as an interesting example from the point of view of the details of the laboratory experiments themselves as well as from the fact that the dimensionless correlations developed on the laboratory scale are also found to hold good on the industrial scale. Only fundamental aspects are discussed in the example and for other details reference must be made to the original paper [9, 10].

Example 3.4 **Mass transfer correlations for gas-stirred slag–metal systems developed from laboratory experiments**

The schematic diagram of the experimental arrangement used is shown in Figure 3.10. The molten copper containing 1 wt % dissolved silicon was oxidized by the liquid $Li_2O–SiO_2–Al_2O_3–FeO$ slag containing 9 wt % FeO. The weights of slag and metal were varied, respectively, as 0.024–0.160 kg and 0.130–2.100 kg. The bath was stirred by injecting argon gas through a nozzle of 0.001 m diameter located at the bottom of the crucible. The effect of the following parameters was studied:

Gas flow rate of injected argon $= 1.80\text{–}6.0 \times 10^{-4}\,m^3/min$

Depth of metal in the crucible $= 0.024\text{–}0.06\,m$

Crucible diameter $\qquad\qquad\quad = 0.034$ and $0.075\,m$

Depth of slag $\qquad\qquad\qquad\quad = 0.012\text{–}0.016\,m$

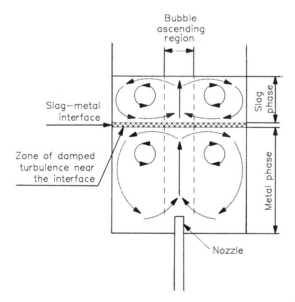

Figure 3.10   Schematic sketch of gas-stirred slag–metal reaction system. (*Source*: Hirasawa, M., Mori, K., Sano, M., Hatanaka, A., Shimatani, J. and Okazaki, Y. (1987) *Trans. Iron Steel Inst. Jpn*, **27**, 277–82, 283–90.)

The similarities with respect to steel refining were as follows:

- The coupled chemical reaction at the interface was essentially the same as the oxidation of silicon in molten iron by FeO in the slag:

$$[Si] + 2(FeO) = (SiO_2) + 2[Fe]$$

- The interfacial tension was of the same order of magnitude as that in molten iron–slag systems.

The mass transfer coefficient was evaluated by plotting the change in concentration of silicon in metal as a function of time and employing the following equation for the metal-phase mass transfer controlled reaction:

$$-\ln\left(\frac{Si_t - Si_e}{Si_0 - Si_e}\right) = k_{Si}\,A\,\frac{\rho_m}{WM}\,t\left(\frac{Si_0}{Si_0 - Si_e}\right) \qquad \text{(A)}$$

where A is the nominal interfacial area, k_{Si} is the mass transfer coefficient of silicon in the metal phase, ρ_m is density of metal, WM is the weight of metal, Si_t is the silicon wt % at time t, Si_e is the equilibrium silicon wt %, and Si_0 is the initial silicon concentration; the derivation of Equation (A) is explained in Example 3.2. The samples of metal were taken at regular intervals and analyzed for silicon. From a plot of $-\ln[(Si_t - Si_e)/(Si_0 - Si_e)]$ versus t the value of k_{Si} was determined at different gas flow rates (Q_g).

The mass transfer coefficient thus determined is plotted as a function of gas flow rate in Figure 3.11. The figure is divided into three regions:

Region I: $Q_g < Q_g^*$; $k_{Si} \propto Q_g^{1/2}$. The superscript * indicates the gas flow rate at the transition point in Figure 3.11 above which the relationship holds.

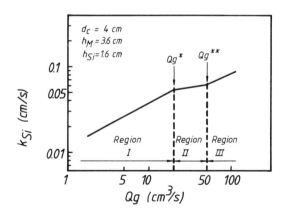

Figure 3.11   Variation of mass transfer coefficient of silicon in metal phase (k_{Si}) as a function of gas flow rate Q_g. (*Source*: Hirasawa, M., Mori, K., Sano, M., Hatanaka, A., Shimatani, J. and Okazaki, Y. (1987) *Trans. Iron Steel Inst. Jpn*, **27**, 277–82, 283–90.)

Region II: $Q_g^* < Q_g < Q_g^{**}$. In this region k_{Si} was nearly independent of Q_g; this was attributed to the development of a zone of damped turbulence near the interface (see Figure 3.10).

Region III: $Q_g > Q_g^{**}$; $k_{Si} \propto Q_g$. The mass transfer coefficient increased linearly with Q_g; this was attributed to mechanical instability at the slag–metal interface due to the passage of gas bubbles, i.e. the ejection of metal droplets into the slag phase or the entrainment of slag droplets in metal.

It is interesting to note the similarity between Figures 3.4 and 3.11. In fact, Figure 3.11 represents a pattern typically observed even in industrial gas-stirred slag–metal systems. The behaviour of regions II and III is still not clearly understood owing to the excessive disturbance of the slag–metal interface caused by the rising gas bubbles. For region I in Figure 3.11, the following correlations in terms of dimensionless numbers were established:

1. For a shallow bath:

$$Sh = C_1 \left[Pe\left(\frac{\rho_m g d_c^2}{\sigma} \right)\left(\frac{h_1}{d_c} \right) \right]^{1/2} \tag{B}$$

2. For a deep bath:

$$Sh = C_2 \left[Pe\left(\frac{\rho_m g d_c^2}{\sigma} \right)\left(\frac{d_b Re^{-n}}{d_c} \right) \right]^{1/2} \tag{C}$$

The different parameters appearing in Equations (B) and (C) are as follows:

- C_1 and C_2 are proportionality constants
- d_c is the diameter of the crucible
- σ is the interfacial tension between slag and metal
- ρ_m is the density of metal
- g is acceleration due to gravity
- h_1 is the distance of the nozzle from the slag–metal interface
- Q_g is the gas flow rate
- d_B is the diameter of the bubble
- Sherwood number, $Sh = k_m d_c / D_m$
- Reynolds number, $Re = [(4Q_g/\pi d_c^2)/v] d_c$
- Péclet number, $Pe = [(4Q_g/\pi d_c^2)/D_m] d_c$

where D_m is the diffusion coefficient of solute in the metal phase, k_m is the mass transfer coefficient in the metal phase and v is the kinematic viscosity.

Knowing the values of C_1, C_2, d_c, d_b, σ, D_m, h_1, Q_g, Sh, Pe, Re and n for a given experimental geometry and the range of gas flow rates studied, it was possible to calculate the value of the mass transfer coefficient. The correlations were found to hold good when extended to ladle desulphurization on an industrial scale. Based on the above laboratory experiments several conclusions were drawn for the optimization of mass transfer in gas-stirred slag–metal systems.

1. Improved mass transfer may be expected for a concentric arrangement of tuyères compared with a linear arrangement.
2. The mass transfer coefficient increases as the depth of the metal phase increases.
3. Mixing time is not a good parameter in predicting mass transfer except when the same tuyère pattern is considered.
4. At high gas flow rates, an increased viscosity of slag is expected to retard the number of droplets ejected from the metal to the slag and thus reduce the effective value of the mass transfer coefficient. A similar influence is expected for interfacial tension. We can stipulate that the steelmaking slags with higher FeO will enhance mass transfer as a result of their lower viscosity as well as lower interfacial tension.
5. A larger volume ratio of slag to metal, up to a certain limit, will enhance the mass transfer coefficient by allowing droplet formation at the slag–metal interface at lower gas flow rates.
6. The effects of gas injection through a submerged lance (up to a depth of 80–90% in metal) or through a tuyère located at the bottom are nearly the same; as the depth of submerged injection is decreased the mass transfer coefficient decreases because less stirring energy is available.

3.6   Gas–metal and gas–slag reactions

The following specific effects at the interface, in addition to mass transfer and chemical reactions, can become rate controlling in gas–metal and gas–slag reactions:

1. Adsorption or desorption   Transfer of a component, to and fro, between gas and slag (or gas and metal) involves the additional steps of adsorption and desorption (see section 3.4). In many cases, depending upon temperature, the composition and physical properties like surface tension of the metal or slag, and the physical conditions of the experiment (like stirring intensity), the adsorption and desorption themselves may become rate controlling.

2. Chemical poisoning   The rate of adsorption (or desorption) at the interface is sometimes found to be very sensitive to the chemical poisoning of the interface by some elements, i.e. surface poisoning of molten iron by oxygen and sulphur.

3. Supersaturation due to delayed nucleation   Slow diffusion of dissolved elements (in metal or slag) and delayed nucleation of gas bubbles due to lack of heterogeneous surfaces may sometimes lead to supersaturation at the interface and hence to a reduced rate of reaction. However, as soon as nucleation sites become available, the reaction occurs at unexpected fast rates to bring down the supersaturation.

The gas–metal and gas–slag reactions of interest to steelmaking are of three types:

1. Oxidation of dissolved impurities in iron like carbon, silicon, manganese and phosphorus, by gaseous oxygen.
2. Reduction of FeO and MnO in slag by rising CO/CO_2 gas bubbles.
3. Dissolution of gases like hydrogen and nitrogen in liquid steel.

The kinetics of oxidation reactions with gaseous oxygen as well as reduction with CO/CO_2 gas mixtures is discussed in detail in Chapter 6 which is devoted to the fundamental refining reactions in primary steelmaking. The decarburization reaction with CO/CO_2 gas mixtures in low-carbon steels is discussed in Chapter 7 which is devoted to the fundamentals of secondary steelmaking. As an example of dissolution reactions, this section is devoted to the kinetics of the dissolution of nitrogen; the kinetics of the dissolution of hydrogen is discussed in Chapter 7.

3.6.1 *Nitrogen adsorption and desorption*

The overall reaction for nitrogen adsorption is

$$N_2(g) \Leftrightarrow 2[N] \quad \text{(dissolved in iron)} \tag{3.106}$$

Various kinetic models proposed for nitrogen have been extensively reviewed [11, 12]. The kinetic steps involved in the transfer of nitrogen from gas to liquid iron are as follows [13]:

1. Transport of nitrogen molecules through the gas phase towards the gas–metal interface.
2. Adsorption of nitrogen at the interface.
3. Chemical reaction at the interface.
4. Transport of nitrogen atoms through the boundary layer in the metal.
5. Transport of nitrogen atoms in the bulk metal.

The steps 1–5 of nitrogen adsorption/desorption and mass transfer can be further broken down into several substeps [11] and any one or a combination of steps may become rate controlling. The oxygen barrier model proposed in 1963 [13] is discussed below. Although the findings of this model were later proved to be incorrect [14], it serves well to explain the fundamental procedure to derive rate equations for nitrogen adsorption/desorption; a similar procedure is followed to derive the rate equations of other models summarized in Appendix B.

Example 3.5 **The oxygen barrier model for the kinetics of nitrogen transfer**

The rate of adsorption and desorption of nitrogen by a clean liquid iron surface in an inductively heated melt has been measured as reported in Pehlke and Elliot [13]. It was noticed that an increase in oxygen and sulphur content of liquid iron led to a drastic decrease in the rate of adsorption (or desorption) of nitrogen.

According to the oxygen barrier model [13], the sequence of steps in nitrogen adsorption are the following:

1. A nitrogen molecule lands on the adsorbed oxygen layer (or adsorbed sulphur layer, as the case may be).
2. The nitrogen molecule dissociates into nitrogen atoms which get adsorbed on the oxygen layer.
3. The nitrogen atoms then jump on to free surface sites and dissolve into liquid iron.

The process of jumping (step 3) is assumed to be slow and reversible:

$$\overset{\displaystyle N}{\underset{\displaystyle O}{|}} + S \Leftrightarrow \overset{\displaystyle N}{\underset{\displaystyle S}{|}} + O$$

where S is a surface site and

$$\overset{\displaystyle N}{\underset{\displaystyle O}{|}}$$

is nitrogen adsorbed on oxygen site. Let C_N^b be the concentration of nitrogen in metal, then

$$- V \frac{dC_N^b}{dt} = A(k_f C_{N-O} C_f - k_b C_{N-S}) \tag{A}$$

where

C_{N-O} = concentration of nitrogen on oxygen

C_{N-S} = concentration of nitrogen on a bare site and bulk

C_f = concentration of free surface sites

V = volume of melt

A = surface area of melt

At high coverages, the fraction of the unoccupied sites is given by the Langmuir adsorption isotherm (discussed in section 3.4.1):

$$(1 - \theta) \propto 1/[O] \tag{B}$$

This is the case of adsorption preceded by dissociation of an oxygen molecule, and from Equation (3.58)

$$(1 - \theta) = \frac{1}{1 + K_e^{1/2} p^{1/2}}$$

Since at high coverages $K_e^{1/2} p^{1/2}$ is much greater than unity

$$(1 - \theta) \propto \left(\frac{1}{p_{O_2}^{1/2}}\right) \propto \left(\frac{1}{[O]}\right) \tag{C}$$

Hence

$$C_f \propto (1 - \theta) \propto 1/[O] = a/[O] \tag{D}$$

where a is a proportionality constant.

Again: C_{N-S} is at equilibrium with nitrogen concentration in the bulk of liquid (C_N^b); C_{N-O} is at equilibrium with nitrogen gas in the atmosphere above the melt.

Since all other steps are assumed to be fast:

$$C_{N-O} \propto p_{N_2}^{1/2} \propto C_e \tag{E}$$

where C_e is the concentration of nitrogen in liquid iron at equilibrium with the atmosphere. Hence

$$- V\frac{dC_N^b}{dt} = A[k_f(a/[O])p_{N_2}^{1/2} - k_b C_{N-S}] \tag{F}$$

where C_{N-S} denotes the concentration of nitrogen per unit area of total surface. The actual concentration at the base sites would be $C_{N-S}/(1 - \theta)$. Because of equilibrium with the bulk concentration

$$\frac{C_{N-S}}{(1 - \theta)} = C_N^b \tag{G}$$

and

$$- V\frac{dC_N^b}{dt} = A(a/[O])[k_f p_{N_2}^{1/2} - k_b C_N^b] \tag{H}$$

At equilibrium

$$p_{N_2}^{1/2} = K_e^{-1} C_e \tag{I}$$

Hence

$$\frac{dC_N^b}{dt} = \frac{1}{V}\frac{dn}{dt} = \frac{A}{V}(a/[O])k_b(C_e - C_N^b) = k(C_e - C_N^b) \tag{J}$$

Integrating Equation (J) gives

$$\log\left(\frac{C_e - C_0}{C_e - C_N^b}\right) = \left(\frac{1}{2.3}\right)\left(\frac{A}{V}\right)\left(\frac{a}{[O]}\right)k_b(t - t_0) \tag{K}$$

where C_0 is the initial concentration and t_0 is initial time. The above equation is of the form

$$\log(C_e - C_N^b) = -At + B \tag{L}$$

The experimentally obtained data fitted well [13] with Equation (L).

In later studies it was found that at high surfactant concentrations, the rate was directly proportional to the nitrogen pressure.

A convincing mechanism for nitrogen adsorption has not been established [12]. Some conclusions of practical importance are the following:

1. The adsorption/desorption rate of nitrogen increases with temperature.
2. The adsorption rate increases with nitrogen partial pressure.
3. In reasonably pure melts at 1873 K the adsorption appears to be a first-order process. At high sulphur and oxygen concentrations both adsorption and desorption rates decrease and the reaction becomes second order.
4. In argon-gas-stirred melts the plume eye can act as a source of nitrogen pickup from the surrounding atmosphere by increasing the exposed surface area by a factor of almost two or more.

3.6.2   *Mass transfer correlations for gas–metal (or gas–slag) systems*

At steelmaking temperatures, there exists only a weak dependence of the mass transfer coefficient on the gas flow rate. The results of some cold model experiments, as well as of bottom blowing of nitrogen in an Fe–C melt, are summarized [1] in Table 3.3 in terms of the effect of the gas flow rate (q) per unit mass of liquid on the capacity coefficient ($\mathcal{K}$). The exponents of q are scattered in the range 0.64 to 1.5. If an attempt is made to separate the capacity coefficient $\mathcal{K}$ as a product of mass transfer coefficient (k) and the interfacial area (A), it is found that the increase in the mass transfer rate is due to an increase in the interfacial area and not because of any direct influence of the gas flow rate (q) on the value of k [1].

Table 3.3   Correlation of mass transfer in gas–liquid systems.

System	Correlation	Remarks	Ref.
water/CO_2		Side blowing	
	$\mathcal{K} \propto q^{1.5}$	$We\,Fr^{-0.4} > 1500$	[1]
	$\mathcal{K} \propto q^{0.7}$	$We\,Fr^{-0.4} < 1500$	
NaOH solution/CO_2	$\mathcal{K} \propto q^{0.7}$	Gas adsorption with chemical reaction	[2]
Water/CO_2	$\mathcal{K} \propto q^{0.64}$		[3]
NaOH solution/CO_2	$\mathcal{K} \propto q$	$A \propto q$	[4]
Fe–C(melt)/N_2	$\mathcal{K} \propto q^{0.65}$	Bottom blowing	[5]

Notes: $\mathcal{K}$, capacity coefficient of mass transfer; q, gas flow rate per unit mass of liquid. For details of each work see [1].

References

1. Asai, S., Kawachi, M. and Muchi, I. (1983) 'Mass transfer rate in ladle refining processes', *Scaninject III, Proc. 3rd Int. Conf. on Injection Metallurgy, Luleå, Sweden, 15–17 June 1983*, Mefos: Luleå, paper 12.
2. Higbie, R. (1935) 'The rate of absorption of pure gas into a still-liquid during short periods of exposure', *Trans. Am. Inst. Chem. Eng.*, **31**, 365–89.
3. Deo, B. and Grieveson, P. (1986) 'Kinetics of desulphurisation of molten pig iron', *Steel Res.*, **57**, 514–19.
4. Swisher, J. H. and Turkdogan, E. T. (1967) 'Decarburization of iron–carbon melts in CO_2–CO atmospheres; Kinetics of gas–metal surface reactions', *Trans. Metall. Soc. AIME*, **239**, 602–10.
5. Robertson, D. G. C., Deo, B. and Ohguchi, S. (1984) 'Multicomponent mixed-transport-control theory for kinetics of coupled slag/metal and slag/metal/gas reactions: application to desulphurization of molten iron', *Ironmaking and Steelmaking*, **11**, 41–55.
6. Ohguchi, S., Robertson, D. G. C., Deo, B., Grieveson, P. and Jeffes, J. H. E. (1984) 'Simultaneous dephosphorization and desulphurization of molten pig iron', *Ironmaking and Steelmaking*, **11**, 202–13.
7. Deo, B., Ranjan, P. and Kumar, A. (1987) 'Mathematical model for computer simulation and control of steelmaking', *Steel Res.*, **58**, 427–31.
8. Robertson, D. G. C. and Staples, B. B. (1974) 'Model studies on mass transfer across a slag–metal interface stirred by bubbles', In *Process Engineering by Pyrometallurgy* (ed. M. J. Jones) Institute of Minerals and Metallurgy: London, pp. 51–9.
9. Hirasawa, M., Mori, K., Sano, M., Hatanaka, A., Shimatani, Y. and Okazaki, Y. (1987) 'Rate of mass transfer between molten slag and metal under gas injection stirring', *Trans. Iron Steel Inst. Jpn*, **27**, 277–82.
10. Hirasawa, M., Mori, K., Sano, M., Shimatani, Y. and Okazaki, Y. (1987) 'Correlation equations for metal-side mass transfer in a slag–metal reaction system with gas injection stirring', *Trans. Iron Steel Inst. Jpn*, **27**, 283–90.
11. Battle, T. P. and Pehlke, R. D. (1986) 'Kinetics of nitrogen absorption/desorption by liquid iron and iron alloys', *Ironmaking and Steelmaking*, **13**, 176–89.
12. Rao, Y. K. and Lee, H. G. (1984) 'Nitrogen absorption and desorption in liquid iron and its alloys', *Trans. ISS*, **4**, 1–10.
13. Pehlke, R. D. and Elliott, J. F. (1963) 'Solubility of nitrogen in liquid iron alloys II. Kinetics', *Trans. Metall. Soc. AIME*, **227**, 844–55.
14. Ito, K., Amano, K. and Sakao, H. (1988) 'Kinetic study on a nitrogen absorption and desorption of molten iron', *Trans. Iron Steel Inst. Jpn*, **28**, 41–8.

4 *Processes and process routes in steelmaking*

4.1 Processes

The technology of steelmaking has evolved through the ages. The ancient men of the Iron Age started to produce iron by heating together lumps of charcoal and iron ore in a pit. Scientific and technological advancements then led to the invention of crucible furnaces, blast furnaces, cupolas, open hearth furnaces, electric furnaces, Bessemer converters, oxygen steelmaking converters, smelting reduction furnaces, ladle furnaces, plasma furnaces, etc. However, unlike other advanced industries, the inventions in the steel industry have been implemented at a slower rate due, primarily, to two reasons. First, owing to the difficulties in carrying out well-controlled experiments at elevated temperatures, the understanding of the fundamentals of the steelmaking process has been far from satisfactory. In fact, until the 1950s, the control of steelmaking processes was regarded as an art, entrusted to an experienced melter. Second, the steel industry is both capital and labour intensive. A steelmaking process, once adopted for bulk production, is used for several years to recover the invested capital. Any new process or process modification, whether in ironmaking or steelmaking, should have a proven success and guaranteed economic benefits to be adopted on a world-wide basis.

Ironmaking and steelmaking have long been accepted as separate disciplines. The end product of ironmaking is molten iron produced through large-capacity blast furnaces or sponge iron produced through smaller-capacity direct or indirect reduction processes. The end product of steelmaking is liquid steel which is delivered to an ingot mould, or to a tundish and then to a continuous casting machine. Looking at the progress made in the last decade in new smelting reduction processes and continuous steelmaking processes, it appears that the distinction between ironmaking and steelmaking will slowly disappear with time.

In between ironmaking and steelmaking there is a narrow but important area of pretreatment of hot metal. The objective of this pretreatment is to control the concentration of dissolved impurities like silicon, phosphorus and sulphur within desired limits before feeding hot metal to the steelmaking shop. In future, and in view of the pace of development of new smelting reduction processes, the intermediate area of hot metal pretreatment may merge with ironmaking. On the other hand, if the continuous steelmaking processes (from ore to steel) become a viable alternative then both ironmaking and steelmaking processes may merge into a single discipline.

Since for the present we agree to accept the separate status of ironmaking and steelmaking, it will be best to associate hot metal pretreatment with ironmaking rather than steelmaking because the pretreatment concerns control of the quality of liquid metal produced from blast furnaces; in principle, almost any quality of hot metal, once brought over to the steelmaking shop, can be refined to a steel of desired composition.

The entire process of steelmaking can be divided into two sequential steps: primary steelmaking in furnaces, and secondary steelmaking (or after treatment) in a ladle. Sometimes, for ordinary grades of steel, the secondary steelmaking step can be bypassed altogether and the metal in the ladle can be sent directly to the tundish or mould for casting.

Operations in the tundish and mould are closely linked to solidification and new names like tundish metallurgy and solidification processing are being coined. Modern developments like thin strip casting and near-net shape casting are linking casting and rolling processes so closely that both solidification and rolling occur simultaneously.

The scope of this book is limited to steelmaking only; a separate treatment of teeming, casting, and solidification and rolling would be more appropriate.

The unit processes of primary and secondary steelmaking are briefly described in sections 4.2 and 4.3, while the fundamentals are discussed in detail in Chapters 5–8. Selection criteria of and typical process routes in steelmaking are given in section 4.4.

4.2   Unit processes for primary steelmaking

The two most important/popular steelmaking processes are the electric arc furnace (EAF) process and the oxygen steelmaking process. The share of EAF in world steel production has steadily increased in the past 30 years and has now reached a level of 35%. Oxygen steelmaking, introduced only in the late 1950s, now accounts for 50–55% of world steel production. Other processes are essentially used to produce special alloy steels.

4.2.1   *The oxygen steelmaking process and its variations*

Oxygen steelmaking is a generic name given to those processes in which gaseous oxygen is used as the primary agent for autothermic generation of heat as a result of the oxidation of dissolved impurities like carbon, silicon, manganese and phosphorus, and to a limited extent the oxidation of iron itself. Several types of oxygen steelmaking processes, like top blowing, bottom blowing and combined blowing, have been invented. A comprehensive overview of the present status of oxygen steelmaking and its expected future trend is provided by Chatterjee *et al.* [1].

The oxygen top-blowing process

In the top-blowing process oxygen is blown with the help of a water-cooled lance inserted through the mouth of the vessel (Figure 2.1). The lance is fitted with a de Laval nozzle at its tip so as to deliver oxygen at supersonic velocity. It may be noted that a supersonic oxygen jet was successfully used first in 1949 in a 2 ton vessel (similar to that in Figure 2.1) at Linz and Donawitz in Austria to refine hot metal (i.e. molten iron produced from a blast furnace, containing carbon, silicon, manganese and phosphorus as impurities). The process was named the 'LD' process after the names of the two places, Linz and Donawitz.

The LD vessel is lined with high-quality basic refractories like tar-bonded dolomite or carbon–magnesite. Due to the fact that basic refractories are used and that oxygen is blown for refining the metal, the LD process is also called the basic oxygen process (BOP) or basic oxygen furnace process (BOF).

The sequence of operations for steelmaking in LD is as follows. The LD vessel is tilted by about 30–40°. Scrap is charged into the vessel (or oxygen steelmaking converter) with the help of a charging bucket. Hot metal is then poured on the scrap. The vessel is then tilted back to its normal upright position for blowing oxygen. An oxygen lance is gradually lowered up to a specified distance from the liquid metal surface (bath surface) and blowing is started simultaneously (while lowering the lance). Within the first 3–6 minutes all the lime is added through mechanized hoppers (or bunkers) to flux the oxides formed by the oxidation of silicon, iron and manganese. Iron ore and spar (CaF_2) may also be added. The lime–silicate slag thus formed essentially contains CaO, SiO_2, MnO, P_2O_5 and FeO. After specified periods the lance is gradually lowered to the lowest predetermined position. Carbon is oxidized to CO/CO_2 (see Chapter 6). The waste gases contain approximately 90% CO and 10% CO_2. Gas recovery hoods are placed on top of the vessel to facilitate the collection of unburnt gases. Total blowing time may vary from 17 to 22 minutes. At the end of the blow the lance is raised. Refined steel is tapped into a ladle and slag is taken out by tilting the converter. Only in the first heat are refractories preheated; thereafter, once the first heat has been tapped, refractories are heated up by iron, slag and gas during the process itself to keep it going during subsequent heats.

The typical end point composition of steel is 0.04–0.06 wt % carbon, 0.2 wt % manganese, 0.02 wt % phosphorus, 0.015 wt % sulphur; the manganese, sulphur and phosphorus contents are a function of input composition of scrap and hot metal. The generation of metal droplets due to the impact of the jet on the metal surface and the evolution of metal and slag composition during the progress of a blow, as well as the kinetics of slag–metal–gas reactions, are discussed in Chapters 5 and 6. The fundamental aspects of supersonic nozzle design are discussed in Chapter 1.

The oxygen bottom-blowing process

As the name suggests, the essential difference from the oxygen top-blowing process (or LD process) is that all of the required oxygen gas is introduced through the bottom of the vessel with the help of tuyères. Lime may be injected in fine powder form along with oxygen (to hasten the dissolution of lime and hence the formation of a well-mixed homogeneous slag). The oxygen bottom-blowing process is also called the OBM, Q-BOP or LWS process depending upon the type of tuyère design used for injecting oxygen through the bottom. Tuyères are shrouded with natural gas or oil in order to protect them from the intense heat generated by oxidation reactions at the tip of the tuyère. The letter 'Q' in Q-BOP stands for 'Quiescent'; compared with the top-blown process, the noise generated and bath eruptions in the Q-BOP are much less. A schematic diagram of the OBM/Q-BOP process is shown in Figure 4.1. The AOD (Argon Oxygen Decarburization) process is also considered by some as a bottom-blowing process but it is used only to produce stainless and special alloy steels.

The combined-blowing process

The term 'combined blowing' implies that gases are blown both from the top and bottom of the converter. A large number of processes have been patented depending

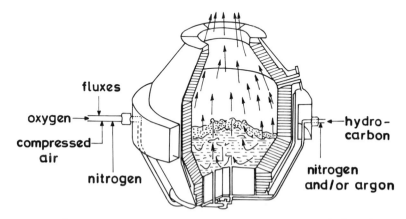

Figure 4.1 Schematic diagram of OBM/Q-BOP process. (*Source*: Brotzmann, K., Kappmeyer, K. K. and Schuerger, T. R. (1976) *Int. Eisenhittentech. Kongr.*, Brussels/Düsseldorf, **1b**, paper 4.1.8)

upon the amount of gas injected from the top and bottom and also the type of tuyère and the gas (or gas mixtures) injected:

- Oxygen top blown plus inert stirring gas injection through the bottom: the processes include LD-KG, LD-AB, LD-BC, LBE, Hoogovens-BAP, etc.; a schematic diagram of LD-KG and LD-AB converters is shown in Figure 4.2.
- Oxygen top blown plus inert and oxidizing gases from the bottom: the processes include LD-CB, LD-OTB, LD-STB, BSC-BAP, etc.; a schematic diagram of LD-OTB and LD-STB converters is shown in Figure 4.3.
- Combined oxygen top- and bottom-blowing processes in which oxygen is injected both from the top and through the bottom: the processes include LD-OB, LD-HC, K-BOP, OBM combined-blowing technique, etc.; a schematic representation of the K-BOP process is shown in Figure 4.4 and a schematic representation of the OBM combined-blowing technique is shown in Figure 4.5.
- Fired blowing processes in which the scrap is first melted with the help of gas (or fuel) combustion through a specially designed lance or burners introduced from the top; hot metal charging may not be required at all and it can be an all-scrap (100%) process to start with. The ratio of scrap to hot metal is decided primarily by relative cost factors; processes include KMS, KS, Krupp–COIN, BSC–carbon injection, etc. A schematic representation of a KMS converter is shown in Figure 4.6.

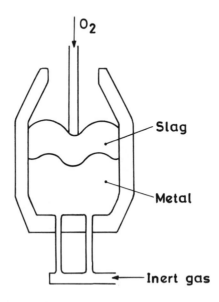

Figure 4.2　Schematic diagram of LD-KG and LD-AB converters. (*Source*: Emi, T. (1980) *Stahl und Eisen,* **100**, 998–1011.)

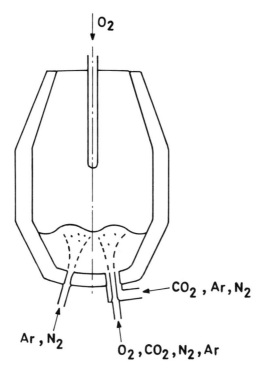

Figure 4.3   Schematic representation of LD-OTB and LD-STB converters showing both injection through single tube and injection through annular tuyère. (*Source*: *Metallurgy of iron*, (1984) Vol. 7: *Practice of Steelmaking*, Springer Verlag: Berlin.)

4.2.2   *The electric arc furnace (EAF) process*

Electricity is the main source of heat generation in the electric arc furnace (EAF). The electrodes are made of graphite and arcing takes place between the metal and the electrodes in the case of direct arc electric furnace.

The main advantage of the arc furnace lies in its flexibility in accepting charge materials in any proportion, namely scrap, molten iron, prereduced material and pellets. The control of electric power can be well exercised to impart heat to the bath at different desired rates. This allows a precise control of refining reactions. Similar to the fired blowing processes, oxygen or fuel can be injected to hasten the processes of melt-down refining. The EAF process thus offers a wide range of possibilities of control to produce ordinary as well as high-quality steels.

A schematic diagram of an EAF is shown in Figure 4.7. The main factor to be considered in the use of EAF compared with the oxygen steelmaking process

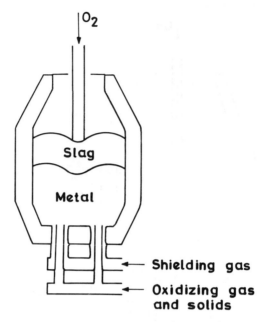

Figure 4.4   Schematic representation of K-BOP process. (*Source*: Emi, T. (1980) *Stahl und Eisen*, **100**, 998–1011.)

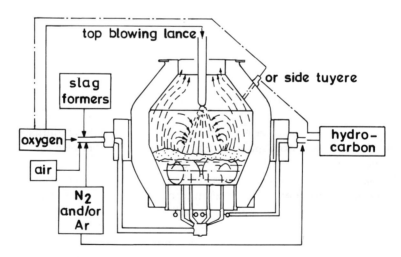

Figure 4.5   Schematic representation of OBM combined-blowing technique. (*Source*: von Bogdandy, L., Brotzmann, K., Fassbinder, H. G., Fritz, E. and Hofer, F. (1982) *Stahl und Eisen*, **102**, 341–6.)

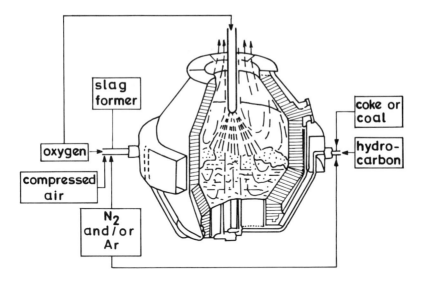

Figure 4.6   Schematic representation of a KMS converter. (*Source*: von Bogdandy, L., Brotzmann, K., Fassbinder, H. G., Fritz, E. and Hofer, F. (1982) *Stahl und Eisen*, **102**, 341–6.)

is the cost of electric power and economy of scale. In some countries electric power is cheap and hence EAF can be used for tonnage production. EAF is best suited for the production of certain special alloy steel grades (namely, tool steels, stainless steel, etc.) because of its inherent advantages of fine process control and the fact that an EAF of almost any size can be made (1 ton to 400 tons); oxygen steelmaking converters of less than 30 tons are not economical because the cost of the oxygen plant has to be considered.

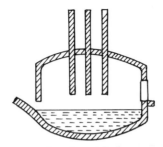

Figure 4.7   Schematic representation of an electric arc furnace (EAF).

4.3   Unit processes for secondary steelmaking

The term 'secondary steelmaking' means the after treatment of the liquid steel produced in a primary steelmaking process. Various possibilities of after treatment of liquid steel, schematically shown in Figure 4.8(a–d) [2–4], can be grouped under the four major headings, namely: stirring processes, injection processes, vacuum processes and heating processes.

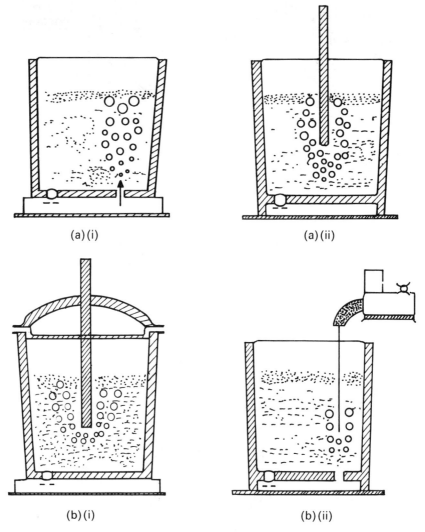

Figure 4.8   Unit processes for secondary steelmaking. (a) Stirring processes: (i) bottom injection; (ii) lance injection. (b) Injection processes: (i) powder injection; (ii) wire feeding.

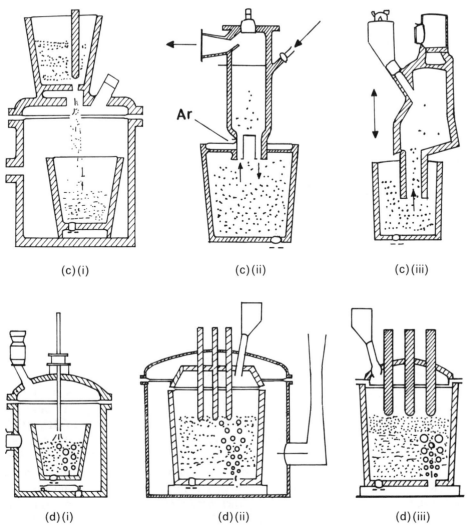

(c)(i)   (c)(ii)   (c)(iii)

(d)(i)   (d)(ii)   (d)(iii)

(c) Vacuum processes: (i) stream degassing; (ii) RH degassing; (iii) DH degassing. (d) Heating processes (with or without vacuum): (i) VOD process; (ii) VAD process; (iii) ladle furnace. (*Sources*: Wiemer, H.-E., Delhey, H.-M., Sperl, M. and Weber, R. A. (1985) *Stahl und Eisen*, **105**, 1142–8; Haastert, H. P. (1983) *Thyssen Tech. Ber.*, **1**, 1–14.)

4.3.1 *Stirring processes*

Stirring (Figure 4.8a) is simple purging of liquid steel by inert argon gas; no specific commercial name has been patented for ladle-stirring processes. Argon can be passed either through a porous plug or through a submerged lance to achieve bath homogenization and oxide flotation. If the ladle is covered with a synthetic slag then the stirring induced by the argon gas also promotes the slag–metal reaction (i.e. desulphurization and deoxidation reactions). The fundamentals of stirring treatment are discussed in Chapter 7.

4.3.2 *Injection processes*

Desulphurization, deoxidation and alloying can be done by introducing the material encased in a wire (called wire feeding) (Figure 4.8b) or by injecting it in a powder form through a submerged lance with the help of a carrier gas, or by injecting argon gas through the bottom of the ladle (stirring process) and laying (dumping) the powder on the top surface of the metal. The fundamentals of injection treatment are discussed in Chapter 8.

4.3.3 *Vacuum processes*

The vacuum processes are designed to reduce the partial pressure of hydrogen, nitrogen and carbon monoxide gas in the ambient atmosphere so that degassing, decarburization and deoxidation can be achieved. In vacuum stream degassing (Figure 4.8c(i)), the liquid metal stream is allowed to fall into an evacuated vessel. In ladle degassing, the ladle is kept in an evacuated vessel and argon is purged through the bottom. Other combinations of simultaneous heating and degassing (VOD, VOR, VODC, VIM, VID, VF, VAD, VODC, CALIDUS in section 4.3.4) are also possible. In recirculation degassing processes, such as in the RH (Figure 4.8c(ii)) and DH (Figure 4.8c(iii)) degassing processes, metal is sucked into an evacuated vessel. In the RH process, for example (Figure 4.8c(ii)), twin snorkel tubes are immersed in the metal. An upward motion of steel is then produced by injecting argon gas in one of the snorkels and applying a vacuum in the vessel at the same time. The liquid steel is sucked into the vessel but leaves by the other snorkel back into the ladle. A continuous recirculation stream of metal between the RH vessel and ladle is set up. Decarburization and degassing occur simultaneously. In the DH process only one snorkel is used and thus, in contrast to the RH process, the DH process operates with repetitive batches of liquid steel being sucked into the vessel and then discharged back to the ladle. As a modification of the RH process, the RH-OB process has been developed in which oxygen can be used for chemical heating and decarburization. The RH and RH-OB processes are more popular than the DH or DH-OB process at present.

The fundamentals of vacuum treatment are described in Chapter 7.

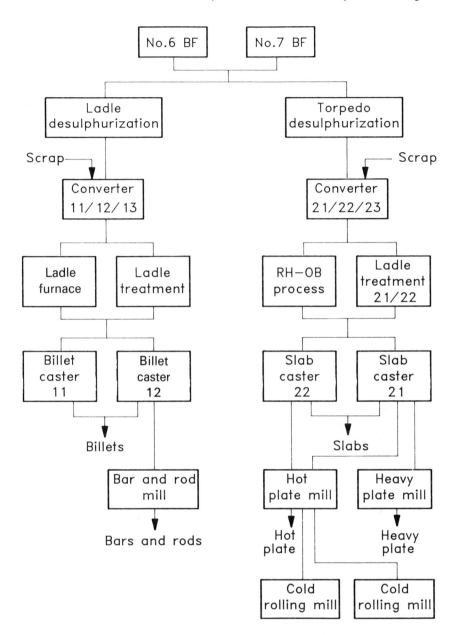

Figure 4.9 Schematic layout of different units in steelmaking at Hoogovens IJmuiden (situation 1991).

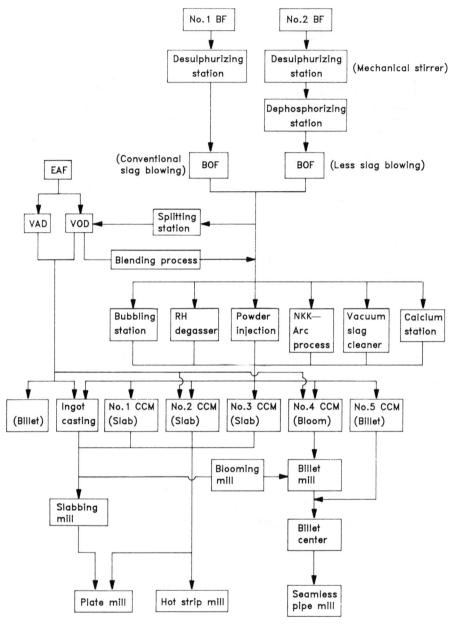

Figure 4.10 Schematic layout of different units in steelmaking at NKK Keihin Works (Japan) (situation 1991)

4.3.4 *Reheating processes*

In all after treatment (or secondary steelmaking) processes, the metal temperature falls due to heat losses to the ambient atmosphere as well as to the heating and melting of the external additions. In order to compensate for this heat loss, either the liquid metal is superheated in the primary steelmaking furnace itself or an additional reheating process is used (Figure 4.8d) such as the ladle furnace (LF) process (with submerged electrodes), the Finkl–Mohr process, the ASEA–SKF process, and the vacuum arc degassing (VAD) process (in which heating is done by submerged electrodes under vacuum to avoid oxidation). On a smaller scale, the vacuum induction melting (VIM), vacuum induction degassing (VID), vacuum induction furnace (VIF) and CALIDUS processes may be used. The CALIDUS process utilizes a specially designed ladle in which the steel is treated, and from which it may be poured. In some processes chemical heating is done by blowing oxygen, for example in the vacuum oxygen refining (VOR) process, vacuum oxygen degassing (VOD) process (see Figure 4.8d), vacuum oxygen decarburization (VODC) process and metal refining process (MRP). Oxygen can also be blown in modified versions of recirculation-type vacuum vessels like RH-OB and DH-OB.

4.4 Selection of process routes in steelmaking

The techno-economic criteria for the selection of an optimum process route (not included in the scope of the present book) depend upon productivity, potential scale of operations, the structure of the cost of a process route, local input prices for raw materials, energy and labour, and also on the ownership patterns (public versus private) and market structure of the particular steel industry concerned [5]. The blast furnace–BOF combination still offers a low-cost route for the production of high-quality steel in integrated steel plants (i.e. steel plants of 1 million ton capacity or more). Two typical examples of the process routes existing at present-day, integrated steel plants are given in Figures 4.9 and 4.10.

References

1. Chatterjee, A., Marique, C. and Nilles, P. (1984) 'Overview of present status of oxygen steelmaking and its expected future trends', *Ironmaking and Steelmaking*, **11**, 117–31.
2. Lange, K. W. (1988) 'Thermodynamic and kinetic aspects of secondary steelmaking processes', *Int. Mater. Rev.*, **33**, 53–89.

3. Haastert, H. P. (1983) 'Pfannenmetallurgishe Verfahren bei der Stahlerzengung' *Thyssen Tech. Ber.*, **1**, 1–14.
4. Cotchen, J. Kevin (1987) 'Ladle metallurgy and the foundry', *Electric Furnace Proc.*, ISS-AIME: New York, pp. 301–9.
5. Aylen, J. (1990) 'Choice of process route in steelmaking', *Ironmaking and Steelmaking*, **17**, 110–17.

5 Primary steelmaking metallurgy: slag formation

5.1 Scope of discussion

The popular phrases 'A good steelmaker is a good slagmaker' or 'Take care of the slag and the metal will take care of itself' are even more pertinent to oxygen steelmaking which uses supersonic oxygen to refine metal within 17–22 minutes; within this short time interval, tons of lime (and dolomite) have to be dissolved and the dynamics of slag formation has to be controlled to arrive at the desired end point temperature and composition of metal. The evolution of the slag and metal compositions during the progress of a blow in oxygen steelmaking is described in section 5.2. The mechanism of lime and dolomite dissolution is discussed in section 5.3.

Besides the addition of fluxing agents like lime, dolomite, ore and fluorspar, the selection of an appropriate nozzle design and the associated blowing regime (i.e. the pattern of variation of lance height and oxygen flow rate during a blow) are the main tools the operator employs to control the behaviour of slag formation. The effect of the decay of the supersonic jet on gas and slag recirculation, metal droplet generation (iron conversion) and impact area (of the jet) is discussed in section 5.4. The interactive effects of nozzle design and blowing regime on mass transfer processes at slag–metal–gas interfaces and hence on slag formation are also described in section 5.4. The foaming of slags is discussed in section 5.5.

5.2 Evolution of slag and metal compositions during the progress of a blow

The chemical compositions of slag and metal, as the blow progresses in an oxygen steelmaking converter, change rapidly with time. Figure 5.1 shows the evolution of bath composition and Figure 5.2 shows the evolution of slag composition in a 300 ton converter [1]. The silicon dissolved in metal is almost completely oxidized in the first 3–4 minutes of the blow. In contrast to silicon, the transfer of sulphur, phosphorus, manganese and carbon takes place over the entire blow period. As the

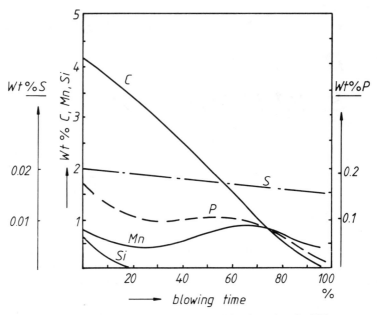

Figure 5.1   Evolution of bath composition with blowing time in 300 ton converter at Hoogovens IJmuiden.

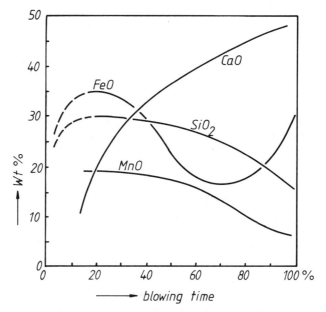

Figure 5.2   Evolution of slag composition with blowing time in 300 ton converter at Hoogovens IJmuiden.

blow progresses, lime (CaO) steadily dissolves in the slag and slag weight increases (Figure 5.3). In the later stages (beyond three-quarters of the blow) the slag weight increases due to lime dissolution as well as to an increase in the iron oxide content of the slag.

The temperature of the liquid metal gradually rises from an initial 1250–1450 °C to about 1600–1680 °C at the end of the blow. The temperature of liquid slag can be higher than that of molten iron by as much as 300 °C in the first few minutes of the blow [2]. The temperature difference between slag and metal gradually reduces to about 50 °C towards the end of the blow. Figure 5.4 shows the liquidus lines in a simple ternary $CaO–FeO–SiO_2$ system (not quasi-ternary) at various temperatures. It may be observed that with increasing temperature the area covered by the liquid region widens, or the solid area covered by $2CaO \cdot SiO_2$ (melting point 2130 °C and henceforth written as C_2S) and $3CaO \cdot SiO_2$ (melting point 2070 °C and henceforth written as C_3S), narrows down.

In order to understand the various compounds and phases that may be present in the slag during converter operation, the chemical compositions in Figure 5.4 can be transposed on to a quasi-ternary diagram of $CaO'–FeO*–SiO'_2$. A typical quasi-ternary diagram on which the actual chemical composition of the slag in a converter has been transposed for different initial silicon contents of hot metal [3] is shown in Figure 5.5; in the case of 0.4% silicon hot metal the slag path enters the C_2S nose earlier than when the silicon content is 0.8%. Each plant has to decide its

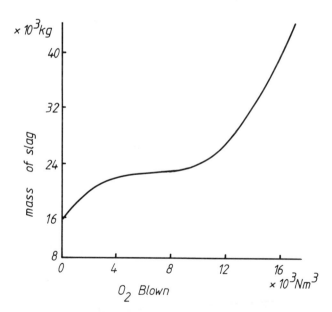

Figure 5.3 Increasing mass of slag with oxygen blown in 300 ton converter at Hoogovens IJmuiden.

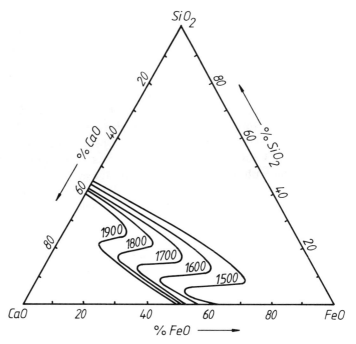

Figure 5.4 Liquidus isotherms in CaO–FeO–SiO$_2$ system at various temperatures.

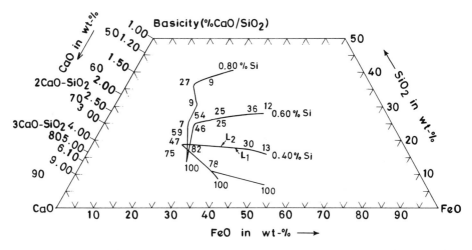

Figure 5.5 Slag formation path with different initial silicon content of hot metal: (a) 0.4%, (b) 0.6%, (c) 0.8%; note that with 0.4% Si the slag formation path is the earliest to enter the C$_2$S nose. (*Source*: Obst, K. H., Schürmann, E., Münchberg, W., Mahn, G. and Nolle, D. (1980) *Arch. Eisenhüttenwes.*, **51**, 407–12.)

own blowing regime and the design of the lance nozzle depending upon the hot metal silicon, size of the converter, hot metal to scrap ratio, lime quality, type of refractory lining, type of fluxing agents used (like CaF_2 and ilmenite), tapping temperatures, cost of raw materials, metallic yield, thermal losses, etc. Considerable research effort, as discussed in section 5.3, has been devoted to understand the mechanisms of the dissolution of calcined lime and dolomite in steelmaking slags with a view to obtaining a good slag formation.

Owing to the non-equilibrium state of reactions in the converter, the behaviour of slag formation is more important than the slag composition at the end of the blow. Often, as a result of differences in the behaviour of slag formation during a blow, two different slag paths may terminate at nearly the same bulk composition in respect of CaO', FeO^* and SiO'_2 in the slag and yet the distribution ratios of phosphorus, manganese and sulphur between slag and metal may not be the same. Practical experience shows that it is important to follow the same behaviour of slag formation and reproduce it from one heat to another in order to obtain consistently good results.

At the beginning of the blow, the distance of the lance tip from the bath surface is kept high (soft blow) and after the formation of an initial slag rich in FeO and SiO_2 (location L_1 in Figure 5.5) the lance is gradually lowered. The slag starts to foam by around one-third to two-fifths of the blow (depending upon the blowing practice) and, as a consequence, the FeO content of the slag begins to decrease towards the precipitation region of dicalcium silicate (C_2S) in the quasi-ternary $CaO'-FeO^*-SiO'_2$ diagram. From about one-half to three-fifths of the blow the foaming is maximum and coincides with the entrance of the slag composition path in the precipitation range of dicalcium silicate (location L_2 in Figure 5.5). Beyond three-quarters of the blow the decarburization rate falls and the FeO content of the slag continuously increases to accommodate most of the oxygen, not used for the oxidation of carbon in the metal, as FeO in the slag. Towards the end of the blow the rate of increase of FeO in the slag depends primarily upon the lance height and the carbon content of the metal and on the amount of undissolved lime and viscosity of the then existing slag.

5.3   Mechanism of lime and dolomite dissolution

The slag produced in the converter in the initial stages, as discussed above, is rich in liquid iron silicate. Calcined lime charged to the converter is almost at room temperature and its dissolution in the slag proceeds in three steps:

1. Slag picks up heat rapidly from the jet impact zone as well as from the coexisting slag and metal phases. Simultaneously the liquid slag penetrates into the pores of the lime; the extent of slag penetration depends upon the size of the pores and the viscosity of the slag.

2. The first chemical attack of the liquid iron silicate $(2FeO \cdot SiO_2)$ or $(Fe_2S)_{(l)}$ slag on solid lime $(C)_{(s)}$ occurs as follows:

$$(Fe_2S)_{(l)} + 4C_{(s)} \rightarrow 2(CF')_{(l)} + (C_2S)_{(s)}$$

The iron-rich calciowustite, $(CF')_{(l)}$, is liquid while dicalcium silicate, $(C_2S)_{(s)}$, is solid at steelmaking temperatures. In porous (soft burnt) lime the dicalcium silicate layer is reported to form at some distance away from the lime [4–6]. In hard burnt lime the dicalcium silicate precipitates first in a granular form and then, due to sintering with calciowustite, it forms a continuous layer around the lime particles which, in turn, hinders further dissolution of the lime. In the case of soft burnt lime particles, the proportion of the FeO-rich liquid $(CF')_{(l)}$ produced is greater than in hard burnt lime. This helps the dicalcium silicate layer in soft burnt lime to rupture earlier than in hard burnt lime.

3. In the later stages of the blow, with the rise in temperature and iron content of the slag, the dicalcium silicate layer softens (due to attack by FeO) and dissolves; the CaO dissolution in the slag abruptly increases.

The equilibrium path of the dissolution of pure solid lime in liquid slag can be traced [7] on the $CaO-FeO^*-SiO_2$ diagram as shown in Figure 5.6. Suppose lime is added to homogeneous slag whose bulk composition is represented by point S.

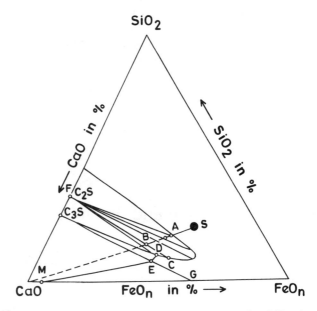

Figure 5.6 Path S–ACDE of dissolution of lime in $CaO-FeO_n-SiO_2$ slag; the path S–ABM indicates ultimate bulk uniform concentration whereas the path S–ACDE indicates the composition of the liquid part of slag. (*Source*: Oeters, F. and Scheel, R. (1974) *Arch. Eisenhüttenwes.*, **45**, 575–80.)

The lime added to the slag has the starting composition corresponding to the point M. The heterogeneous mixture of solid lime and liquid slag begins to react and only if all the lime dissolves immediately would the bulk slag composition, depending upon the amount of lime added, change along the path SABM; the liquid part of the slag would follow the contour ACDE. For example, at point B the ratio of solid slag (C_2S) to liquid slag (of composition C) will eventually be BC/BF. The lime, however, does not dissolve instantaneously because the first reaction product formed on lime is C_2S (corresponding to the point A). C_2S has a high melting point and forms, as explained above, a layer on the undissolved lime particles. Further dissolution of lime captured underneath the C_2S layer is prevented until favourable conditions arise which dissolve or rupture the C_2S layer itself. Along the surface DE the liquid slag is saturated with C_3S and on the surface GE it is in equilibrium with a solid solution of $CaO-FeO-SiO_2$. If the initial composition of liquid slag is close to the surface DE then C_3S should form first but the kinetics of precipitation of C_3S is extremely slow. As a result, C_2S generally forms first followed by precipitation of C_3S within or underneath the layer of C_2S at a much later stage.

Relatively recent studies [4] have shown that in the presence of dolomite (i.e. 5% MgO in the slag), the dissolution mechanism of lime changes in such a way that the dissolution rate is slowed down as a result of the early precipitation of tricalcium silicate next to solid lime:

Iron silicate (l) + wustite dendrites	granular dicalcium silicate	tricalcium silicate (s) + FeO-rich liquid (l)	lime (s) + FeO-rich liquid

The dissolution of dolomite $(CM)_{(s)}$ proceeds according to the reaction

$$(Fe_2S)_{(l)} + 2(CM)_{(s)} \rightarrow (C_2S)_{(s)} + 2(MF')_{(s)}$$

where $(MF')_{(s)}$ represents the solid solution of MgO and FeO, i.e. magnesiowustite. The reaction interfaces formed in the initial stages may be represented as

Iron silicate melt (l)	wustite (s) + melt (l)	zone of MgO–FeO enrichment (s + l)	dicalcium silicate (s) + magnesio- wustite (MF')	unreacted dolomitic lime

The magnesiowustite has a high melting point and since it forms a continuous layer on the dolomite particles (thereby closing all the pores) the dissolution rate of dolomite is slowed down in a similar way as that of lime by the dicalcium silicate layer.

The rate of dissolution of dolomite particles in steelmaking slags is known to be slower than that of lime. A recent study [8] has shown that if both lime and dolomite are added together the proportionate dissolution of calcined dolomite (expressed as the weight of dolomite dissolved divided by the weight of dolomite added) is about half the proportionate dissolution of lime (the weight of lime dissolved divided by the weight of lime added) at about one-quarter of the blowing time. No convincing explanation has yet been offered for this phenomenon.

5.4 Selection of nozzle design and blowing regime to control slag formation

The selection of a suitable combination of nozzle design and the associated variation of lance height pattern during a blow (i.e. blowing regime) is crucial for the control of slag formation in oxygen steelmaking converters. The dynamics of slag formation in the converter is directly controlled by the jet decay characteristics, the metal droplet generation rate (or iron conversion), the recirculation of metal, slag and gas, the impact depth of the jet and the impact area of the jet on the bath surface. The design features of de Laval nozzles used at some plants are summarized in Table 5.1. The blowing regimes used (corresponding to the nozzle designs in Table 5.1) are schematically shown in Figure 5.7.

The fundamentals of nozzle design for top-blown oxygen steelmaking converters have been reviewed [9] and empirical correlations have been proposed for: mass flow rate versus converter capacity, upstream pressure versus converter capacity, total throat diameter versus converter capacity, aspect ratio versus converter capacity,

Table 5.1 Lance head design data of some steel plants.

		Plant 1 #31	Plant 1 #37	Plant 1 #02	Plant 2	Plant 3	Plant 4	Plant 5	Plant 6
F_{O_2}	(m³/min)	915	915	660	650	750	900	900	330
p_o	(Pa)	10.2	10.8	6.6	11.1	13.3	11.6	11.0	7.9
d_t	(mm)	46.6	45.2	55.0	37.6	36.9	48.44	40.6	41.0
d_e	(mm)	58.0	61.3	65.75	55.35	51.95	63.0	57.75	50.5
θ_1	(deg)	12	14	10	14	16	12	14	10
n		5	5	4	5	5	4	6	3
W_{bath}	(tons)	315	315	315	245	230	330	300	100
Bottom-stirring system		BAP	BAP	BAP	tuyères ϕ 5.5 mm	LBE	BAP	–	BAP
$Si_{hot\,metal}$	(%)	0.400	0.400	0.400	0.330	0.400	0.750	0.600	0.400

Notes: Plant information:
  Plant 1: Hoogovens IJmuiden, BOS No. 2,
      #31: lance head design 31, converter 23;
      #37: lance head design 37, converter 23;
      #02: lance head design 02, converter 21 and 22.
  Plant 2: BS Lackenby.
  Plant 3: Sollac Dunkerque.
  Plant 4: BS Port Talbot.
  Plant 5: BS Scunthorpe.
  Plant 6: Hoogovens IJmuiden, BOS No. 1.

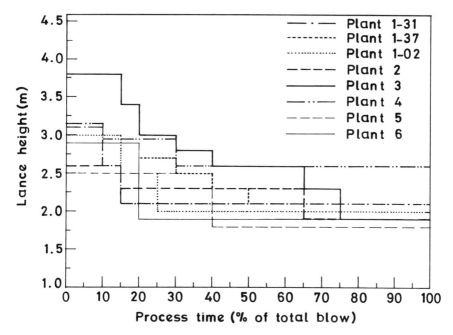

Figure 5.7   Blowing regime of various plants; for details of plant names see footnote to Table 5.1.

specific oxygen blowing rate versus converter capacity, bath diameter versus converter capacity, and initial lance distance versus converter bath diameter. These empirical correlations can provide only an approximate idea of the nozzle dimensions to be used. The selection of the actual dimensions of a nozzle and the associated blowing regime depends upon specific slag formation conditions existing at a particular plant.

The criteria used in the selection of a nozzle design and blowing regime with particular reference to slag formation are discussed in this section. A new concept of controlling the dynamics of slag formation through a dimensionless 'droplet ejection and oxidation' number (or DEO number) is derived from first principles and applied to 100 and 300 ton converters.

5.4.1   *Decay of supersonic jet due to entrainment and its effect on gas and slag recirculation*

Given the density of oxygen gas as 1.429 kg/m³, the oxygen flow rate (F_{O_2}) through a lance tip fitted with n nozzles of throat diameter d_t (m) can be calculated from

$$F_{O_2} = 1.414 \times 10^5 \xi \, \frac{n d_t^2 p_o}{\sqrt{T_o}} \; (\text{m}^3/\text{min}) \tag{5.1}$$

where ξ, the coefficient of flow rate, lies in the range 0.95–0.97, p_o is absolute pressure (in pascals) and T_o is temperature (in kelvin) at the nozzle entrance.

The supersonic jet issuing from the nozzle has a supersonic core in which the gas velocity is higher than Mach 1. The diameter of the core gradually narrows down with increasing distance from the lance tip. For the nozzles used in oxygen steelmaking, the length of the supersonic core usually varies from three to eight times the nozzle exit diameter.

After the decay of the supersonic core, the jet begins to spread at an included angle of approximately 15° due to entrainment of gases from the ambient atmosphere. The entrainment has at least three physical effects: (a) the mass of the jet increases, (b) the velocity of the jet decreases and (c) if the gases in the atmosphere are at a higher temperature than the jet, then the jet temperature increases. The atmosphere inside the steelmaking converter consists mainly of carbon monoxide gas in the temperature range 1773–1993 K. The oxygen jet can therefore be visualized as a flame of oxygen [10] burning in an atmosphere of carbon monoxide gas.

The relationship between the jet centreline velocity $(u_{o,x})$ at any distance x from the lance tip and the velocity $(u_{r,x})$ at radius r and distance x from the lance tip can be expressed in the functional form

$$\frac{(u_{r,x})}{(u_{o,x})} = f\left(\frac{r}{x}\right) \tag{5.2}$$

If a Gaussian distribution of the velocity is assumed, then

$$\frac{(u_{r,x})}{(u_{o,x})} = \exp\left[-\frac{1}{2c}\left(\frac{r}{x}\right)^2\right] \tag{5.3}$$

where c is a constant, called in the present work a nozzle constant. The ratio of the kinetic energy of the jet at (r, x) and (o, x) can be written as

$$\frac{\rho_x(u_{r,x})^2}{\rho_o(u_{o,x})^2} = \exp\left[-\frac{1}{c}\left(\frac{r}{x}\right)^2\right] \tag{5.4}$$

where ρ refers to density.

From conservation of energy at the nozzle exit $(x = 0)$ and at any distance x from the nozzle exit

$$\int_0^\infty \rho_x u_{(r,x)}^2 2\pi r \, dr = \tfrac{1}{4}\pi d_e^2 \rho_e u_e^2 \tag{5.5}$$

where d is the diameter and the subscript e refers to the conditions at the nozzle exit. Substituting for $u_{r,x}$ from Equation (5.3) and also the value of the integral

$$\int_0^\infty \exp(-x) \, dx = 1$$

we get from Equations (5.4) and (5.5)

$$\frac{\rho_x u_{o,x}^2}{\rho_e u_e^2} = \frac{1}{4}\left(\frac{d_e}{x}\right)^2 \frac{1}{c} \tag{5.6}$$

In the presence of the supersonic core, the entrainment of the gases from the ambient atmosphere is negligible, and hence

$$\frac{\rho_x u_{o,x}^2}{\rho_e u_e^2} = 1 \tag{5.7}$$

Let the length of the supersonic core be x_{ss}. On substituting Equation (5.7) into Equation (5.6), we get

$$x_{ss} = \frac{d_e}{2\sqrt{c}} \tag{5.8}$$

The volumetric recirculation of the ambient gas due to jet entrainment is defined as the ratio of the volume flow rate of the gas (ϕ_x) at any distance x to the volume flow rate of gas (ϕ_e) at the nozzle exit. The value of ϕ_x can be calculated from

$$\phi_x = 2\pi \int_0^\infty u_{r,x} r \, dr \tag{5.9}$$

On substituting $u_{r,x}$ from Equation (5.3):

$$\phi_x = 2\pi c x^2 u_{o,x} \int_0^\infty \exp\left[-\frac{1}{2c}\left(\frac{r}{x}\right)^2\right] d\left[\frac{1}{2c}\left(\frac{r}{x}\right)^2\right]$$

$$= 2\pi c x^2 u_{o,x} \tag{5.10}$$

The dynamic pressure ($p_{d,x}$) at any distance x on the jet centreline is defined as

$$p_{d,x} = \tfrac{1}{2}\rho_x u_{o,x}^2 \tag{5.11}$$

Similarly, the dynamic pressure at the nozzle exit is

$$p_{d,e} = \tfrac{1}{2}\rho_e u_e^2 \tag{5.12}$$

From Equations (5.10)–(5.12) we get

$$\phi_x = 8c\left(\frac{x}{d_e}\right)^2 \left(\frac{\rho_e}{\rho_x}\right)^{1/2} \left(\frac{p_{d,x}}{p_{d,e}}\right)^{1/2} \phi_e \tag{5.13}$$

It follows from Equations (5.6), (5.11) and (5.12) that

$$c = \frac{1}{4}\left(\frac{d_e}{x}\right)^2 \frac{p_{d,e}}{p_{d,x}} \tag{5.14}$$

Evaluation of nozzle constant c

The value of the nozzle constant c in Equation (5.14) has been determined at Hoogovens IJmuiden by carrying out experiments with freely expanding jets and evaluating the data as follows. Inside the nozzle, the static pressure can be substituted for dynamic pressure, i.e. $p_{d,e} = p_o - p_e$. Hence the ratio of dynamic pressure at any distance x and the pressure at the nozzle exit can be written as

$$\frac{p_{d,x}}{p_{d,e}} = \frac{p_{d,x}}{p_o - p_e} = \frac{p_{d,x}}{p_o}\left(\frac{1}{1 - p_e/p_o}\right) \tag{5.15}$$

The ratio d_e/x can be expressed as

$$\frac{d_e}{x} = \left(\frac{d_t}{x}\right)\left(\frac{d_e}{d_t}\right) \tag{5.16}$$

where d_t is the throat diameter. From Equations (5.14)–(5.16)

$$\frac{p_{d,x}}{p_o - p_e} = \frac{p_{d,x}}{p_o(1 - p_e/p_o)} = \frac{1}{4c}\left(\frac{x}{d_t}\right)^{-2}\left(\frac{d_e}{d_t}\right)^2 \tag{5.17}$$

For any given nozzle, the value of c can be experimentally determined by measuring $p_{d,x}$ as a function of the distance x from the nozzle exit and then plotting $p_{d,x}/(p_o - p_e)$ or $p_{d,x}/p_o$ versus x/d_t. Experiments have been carried out at Hoogovens to determine these plots for Mach numbers ranging from 1.7 to 2.2. According to Equation (5.17), the theoretical value of the slope from a log–log plot is -2 but from experiments the value of the slope is found to depend upon exit pressure (p_e); when the exit pressure is approximately 1 atm the value of the slope is less than -2. For example, in the case of a Mach 2 nozzle the experimental relationship between $p_{d,x}/p_o$ and x/d_t is

$$\frac{p_{d,x}}{p_o} = 230\left(\frac{x}{d_t}\right)^{-2.4} \tag{5.18}$$

Let us take a typical example of a nozzle with the following specifications: $p_o = 10.6$ atm, $Ma = 2.1$, $d_t = 0.0452$ m, $d_e = 0.0613$ m and $p_e = 1$ atm; then the value of c is 6.6×10^{-3}. On substituting the value of c and d_e in Equation (5.8), the length of the supersonic core, x_{ss}, can be calculated (0.49 m). The length x_{ss} is thus approximately eight times the nozzle exit diameter, $d_e(= 0.0613$ m).

Evaluation of gas entrainment ratio

On substituting the nozzle constant c from Equation (5.14) into Equation (5.13)

$$\frac{\phi_x}{\phi_e} = 2\left(\frac{\rho_e}{\rho_x}\right)^{1/2}\left(\frac{p_{d,e}}{p_{d,x}}\right)^{1/2} \tag{5.19}$$

The ratio of the mass stream (ϕ'_x) of the gas at distance x to the mass stream of the gas at the nozzle exit, expressed as R_g, can then be written as

$$R_g = \frac{\phi'_x}{\phi'_e} = 2\left(\frac{\rho_x}{\rho_e}\right)^{1/2}\left(\frac{p_{d,e}}{p_{d,x}}\right)^{1/2} \tag{5.20}$$

The density of gas at the exit can be calculated from

$$\rho_e = \frac{p_e}{p_s}\frac{T_s}{T_e}\rho_s \tag{5.21}$$

where p is pressure and the subscript s refers to standard conditions. Since the expansion of the gas is isentropic, the ratio of inlet temperature (T_o) to exit temperature (T_e) can be expressed as

$$\frac{T_o}{T_e} = \frac{T_s}{T_e} = \left(\frac{p_o}{p_e}\right)^{(\gamma-1)/\gamma} \tag{5.22}$$

The density of gas at distance x from the nozzle depends upon local conditions of temperature and pressure; similar to Equation (5.21)

$$\rho_x = \frac{p_x}{p_s}\frac{T_s}{T_x}\rho_s \tag{5.23}$$

Substitution of Equations (5.21)–(5.23) into Equation (5.20) gives

$$R_g = \frac{\phi'_x}{\phi'_e} = 2\left(\frac{p_x}{p_e}\right)^{1/2}\left(\frac{T_s}{T_x}\right)^{1/2}\left(\frac{p_e}{p_o}\right)^{(\gamma-1)/2\gamma}\left(\frac{p_o - p_e}{p_{d,x}}\right)^{1/2} \tag{5.24}$$

For a Mach 2.1 nozzle, substitution of Equation (5.18) into Equation (5.24) gives

$$R_g = \frac{\phi'_x}{\phi'_e} = 0.132\left[\left(\frac{p_x}{p_e}\right)\left(\frac{T_s}{T_x}\right)\left(\frac{p_e}{p_o}\right)^{(\gamma-1)/\gamma}\left(1 - \frac{p_e}{p_o}\right)\right]^{1/2}\left(\frac{x}{d_t}\right)^{1.2} \tag{5.25}$$

The value of R_g is plotted as a function of lance height (x) for the nozzle CONV23/37 in Figure 5.8; as expected, R_g increases with lance height.

The nozzles used in oxygen steelmaking are so designed that the exit pressure is approximately equal to or slightly higher than the ambient pressure. Otherwise, shock waves are generated. If the exit pressure is lower than the ambient pressure then there will be a tendency for the dust-laden gases to be sucked into the nozzle from the outer periphery of the nozzle wall, thereby causing enhanced nozzle wear. Experiments at Hoogovens have shown that the lower critical limit of exit pressure p_e is: $p_e \geqslant 0.5 \times$ ambient pressure. The pressure of the gases inside the converter is approximately 1 atm except during the peak decarburization period when the pressure can rise to approximately 1.15 to 1.20 atm. The effects of pressure changes in the ambient atmosphere on nozzle performance are discussed by Inada and Watanabe [11].

An additional design requirement for the multihole lance tips is that the jet produced from each nozzle should not interfere with jets from other nozzles (i.e.

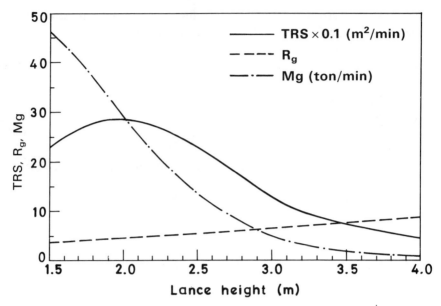

Figure 5.8 Variation of total reaction surface (TRS), gas recirculation (R_g) and iron conversion (Mg) with lance height; for details of lance head 37 see Table 5.1.

non-coalescing jets). This is achieved by placing each nozzle at an angle from the vertical axis. Practical experience at Hoogovens has confirmed that in order to avoid jet coalescence the distance between the nozzle centres should be greater than $2.5d_t$ (where d_t is the throat diameter).

The conditions inside the converter are not ideal for the theoretical relations derived for the freely expanding jet to be completely valid. For example, in the case of a converter the metal (outside the jet impact zone) is covered with a layer of slag; the jet is confined in a vessel from which waste gases $(CO + CO_2)$ are emerging; and both gas temperature and composition vary along the axis of the jet due to entrainment. It is therefore necessary to fine-tune the fundamental relations developed above on the basis of actual results.

The recirculation of the gas in the converter is directly proportional to R_g. If the metal is covered by a layer of slag then the recirculation of slag will also be proportional to R_g. At a given lance height the extent of slag and gas recirculation will depend upon the height of the slag and the nature of the slag, i.e. foamy slag, viscous slag or thick slag. In principle the length of the supersonic core and slag thickness should be subtracted from the lance height in calculating R_g (the entrainment of atmospheric gases within the length of the supersonic core of the jet is negligible). Several empirical relations have been proposed to predict the length of the supersonic core [12] but the predicted length is approximately the same as that calculated from Equation (5.8).

5.4.2   *Specific reaction surface of metal droplets, iron conversion and depth of jet penetration due to impingement*

The metal droplets are generated as a result of jet impact and the shearing action of the gas flow from the impact region when the jet strikes the metal surface and the gases are deflected upwards. It has been found that the drop size distribution obeys the Rosin–Rammler–Sperling (RRS) distribution function [13]

$$RRS = 100 \exp[-(d/d')^n] \quad (\%) \tag{5.26}$$

where d is droplet diameter, d' is a measure of fineness of particle (the reciprocal of d' characterizes the degree of drop break-up) and n is a measure of homogeneity of the particle size distribution; it is independent of d'.

The average value of n for hot metal and steel is 1.26. The drop size distribution (RF, in fractions) over a wide range of blowing conditions is given by

$$RF = (0.001)^{(d/d_{\text{limit}})^{1.26}} \tag{5.27}$$

where d_{limit} is the value of d corresponding to RRS = 0.1%.

From Equation (5.27) the fraction of droplets with diameter d, $f(d)$, is given by

$$f(d) = 8.7038RF \frac{d^{0.26}}{d_{\text{limit}}^{1.26}} \tag{5.28}$$

For a non-coalescing jet the value of d_{limit} (in mm) is related to different parameters [14] as follows:

$$d_{\text{limit}} = 5.513 \times 10^{-3} \times \{10^6(d_t^2/x^2)p_a[1.27(p_o/p_a) - 1]\cos \theta\}^{1.206} \tag{5.29}$$

where

d_t = throat diameter, m

p_o = supply pressure, atm

p_a = atmospheric pressure, atm

x = height of the nozzle from the metal surface, m

θ = angle of the divergent part of the nozzle.

The specific reaction surface (SRS) of the droplets can be calculated from

$$SRS = \left(\int_0^{d_{\text{limit}}} n f(d) \pi d^2 d(d) \right) \Big/ (\text{mass of droplets}) \ (\text{m}^2/\text{kg}) \tag{5.30}$$

where n is the total number of droplets produced. SRS increases with lance height. The total mass (ton) of metal droplets ejected per minute (Mg) due to the impingement of an oxygen jet on the metal surface is called iron conversion (tons/min). Schoop *et al.* [15] proposed the following empirical relationship between the metal droplet stream, Mg, and the specific blowing impact at a lance height x,

I_x:

$$Mg = -223.067 + 76.632 \log I_x \quad \text{(tons/min)} \tag{5.31}$$

The specific blowing impact, I_x, is given by

$$I_x = 10^4 A_{tot} (0.155 F_{O_2} - 9.81)/x^2$$

where

$A_{tot} =$ sum of nozzle areas at exit, m^2

$x =$ lance height, m

$F_{O_2} =$ oxygen flow rate, m^3 (stp)/min

The constants appearing in Equation (5.31) should be tuned to the specific situation at a particular plant; for example, when Equation (5.31) is used as such for 300 ton converters at Hoogovens, a negative value of iron conversion is obtained at a lance height greater than 3 m. Also, for smaller capacity (i.e. 100 ton) converters, Equation (5.31) is found to predict excessively high values of iron conversion (in comparison with 300 ton converters).

On the basis of water model experiments and cold model experiments employing liquid mercury, He and Standish [16] have proposed a functional relationship between iron conversion and Weber number (We):

$$Mg = f(We) \tag{5.32}$$

where the Weber number is defined as

$$We = \frac{\rho_g u_{o,x}^2}{(\rho_e g \sigma_e)^{1/2}} = \frac{2 p_{d,x}}{(\rho_e g \sigma_e)^{1/2}} \tag{5.33}$$

The subscript l refers to liquid metal.

In order to apply Equation (5.32) to real systems, the droplet generation per unit volume of blown gas must be considered. The plot of Mg/F_{O_2} versus Weber number is shown in Figure 5.9. The suitability of Equation (5.32) for top-blown converters can be tested as follows.

According to Figure 5.9, iron conversion is negligible when the Weber number is approximately 10. On substituting the values for liquid steel, $We = 10$, $\sigma_e = 1.7$ N/m, $\rho_g = 1.36$ kg/m^3 and $g = 9.81$ m/s^2, in Equation (5.33), the jet centreline velocity at the metal surface ($u_{o,x}$) can be calculated as 50 m/s. The dynamic impact pressure of the jet on the metal surface can then be used to calculate the depression (or jet penetration), Δx, in the metal bath:

$$p_{d,x} = \tfrac{1}{2} \rho_g u_{o,x}^2 = \rho_l g \Delta x \tag{5.34}$$

Several empirical relations have been proposed in the literature [17–20] (see Table 5.2) to calculate the depth of jet penetration (Δh) in a liquid metal bath. When the relations in Table 5.2 are applied to a 300 ton converter, the predicted values of Δx vary only in a small range provided the lance height is greater than 2.5 m (Figure

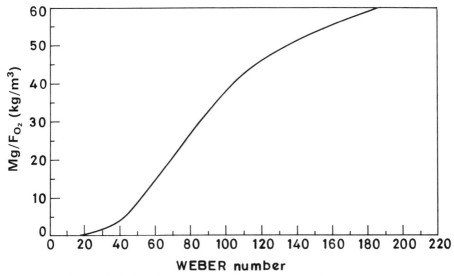

Figure 5.9 Specific iron conversion versus Weber number.

5.10); the line corresponding to [18] or [19] in Figure 5.10 can be used to predict an average value of the depth of jet penetration. At lance heights lower than 2 m, the differences in the predicted values of jet penetration are large (Figure 5.10). This may partly be attributed to the experimental difficulties involved in the accurate measurement of the depth of jet penetration. At small lance heights (less than 2.5 m), it is observed experimentally that the cavity formed on the liquid metal surface

Table 5.2 Calculation of impact depth.

Ref. [17]	$X = \ln(M/h^3)$ $\ln(\Delta_h/h) = -0.041\,47 \times X^2 + 1.3418 \times X - 8.4297$
Ref. [18]	$M_h = \dfrac{0.7854 \times d_t^2 \times p_{amb}(1.27 \times p_o/p_{amb} - 1)\cos(\alpha)}{g\rho_1 h^3}$ $\Delta_h/h = 4.469 \times M_h^{0.66}$
Ref. [19]	$X = 7 \times d_e$ $\Delta_h/h \times [1 + (\Delta_h - X)/h]^2 = 100 \times M/(\pi\rho_1 gh^3)$
Ref. [20]	$\ln[pc/(p_o - p_{amb})] = -2.417 \times \ln(h/d_t) + 5.8307$ $\Delta_h = p_{d,h}/(\rho_1 g)$
Hoogovens	$\ln(pd/h) = -2.3654 \times \ln(h/dt) + 5.4402$ $\Delta_h = p_{d,h}/(\rho_1 g)$

Note: $M = 1.421 \times dt^2 \times p_o[1 - (p_{amb}/p_o)^{0.286}]^{1/2}$; Δ_h, impact depth (m); h, nozzle–bath distance (m); ρ_1, density of steel (kg/m^3); p_{amb}, ambient pressure (Pa abs.); p_o, pressure before nozzle (Pa abs); g, acceleration due to gravity (m/s^2); d_t, throat diameter of nozzle (m); d_e, exit diameter of nozzle (m); α, angle between nozzle and lance axis (deg); $p_{d,h}$ is dynamic impact pressure at lance height h.

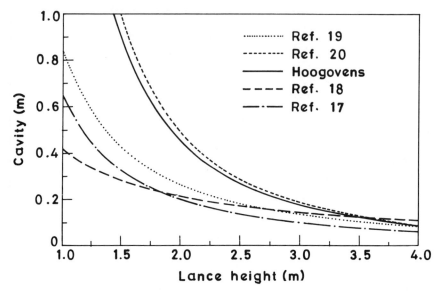

Figure 5.10   Cavity depth due to impingement of jet on bare metal surface; the formulae used are given in Table 5.2.

becomes unstable due to the increased turbulence energy of the bath and also to the entrainment of large metal droplets in the jet which fall back to the metal.

On substituting the velocity $u_{o,x} = 50$ m/s in Equation (5.34), the value of Δx is found to be 0.025 m, i.e. iron conversion is nearly zero when the jet penetration is 0.025 m or less. The experiments carried out by Molloy [21] have also shown that a transition from penetrating jet to dimpling jet occurs when jet penetration is less than 0.03 m. On the basis of present work the critical jet centreline velocity ($u_{o,x}$) for a transition from dimpling jet to splashing jet may be taken as approximately 50 m/s.

The calculated values of iron conversion, for the nozzle design 37 (mentioned in Table 5.1), are plotted as a function of lance height in Figure 5.8. The total reaction surface generated per minute (TRS = SRS × Mg, m²/min) is also plotted in Figure 5.8; it shows a maximum at a lance height of 2.05 m.

5.4.3   *Impact area of the jet*

The jet impact area on a metal bath can be calculated in two ways for a multihole lance: the area of individual impact spots corresponding to each nozzle can be added to give the total area; alternatively, one may join the outer periphery of all the impact spots by an enveloping circle (say of diameter I_o) and calculate the area of this circle. If the diameter of one spot is I_A then it can be shown that the value of I_o/I_A will be 2 or less for coalesced jets.

In order to calculate I_A, the periphery of the jet impact spot is defined up to the point of 3 mm depression in the molten iron surface. Let θ_1 be the nozzle angle and θ_2 the half angle of the cone formed by the jet boundary. If θ_2 is much smaller than θ_1 then as a first approximation

$$\tan(\theta_1 + \theta_2) = \tan \theta_1 + \tan \theta_2 \tag{5.35}$$

The diameter I_A of an impact spot will then be given by

$$I_A = 2(\tan \theta_1 + \tan \theta_2)x \tag{5.36}$$

where x, as before, is the lance height. The value of $\tan \theta_2$ is related to the maximum depth of depression of metal in the impact spot (Δx):

$$(\tan \theta_2)^2 = -[3 + \lg(0.003c/\Delta x)] \tag{5.37}$$

where c is the nozzle constant defined in Equation (5.3) and 0.003 m is the minimum depression in the metal surface at the jet boundary. In actual plant practice an adapted expression is employed to calculate Δx depending upon slag volume and converter geometry. The total impact area is calculated by multiplying the area of one spot by the number of nozzles. The impact area increases with θ_1. The inclination angle of the nozzle is increased with the number of nozzles to avoid jet coalescence [22].

5.4.4 Effect on slag formation of mass transfer processes at the interfaces of jet, gas, metal droplets, slag and bulk metal

We know that vigorous chemical reactions take place on the surface of metal droplets ejected from the jet impact zone because all the oxygen supplied by the jet is consumed immediately in oxidation reactions. Carbon, silicon, manganese, phosphorus and sulphur present on the surface of the droplets are oxidized and the excess oxygen is converted to FeO. The surface of the metal droplets ejected from the jet impact zone is thus covered with a layer of FeO. The activity of FeO around the metal droplet is greater than in the bulk slag. The ejected droplets are caught up in the slag which is present around each jet. Some of the droplets may be entrained into the neighbouring jets (in the case of a multihole nozzle) and be reoxidized. This may happen, for example, in the initial stages of the blow period when the amount of slag is small. The FeO produced, besides dissolving the lime, attacks the refractory lining. Considerable oxidation of the carbon monoxide formed in the jet impact zone to carbon dioxide (postcombustion of the gases) takes place in the initial stages of the blow leading to a high carbon dioxide concentration in the exhaust gas (see Chapter 6) even before being caught up in the slag. However, with time, as the slag volume and the foaming of slag increase, the postcombustion decreases.

During the main part of the blow the lance becomes submerged in the slag–metal–gas foam or emulsion. A part of the foam (or emulsion) is sucked into the jet as a result of the entrainment effect. A thorough mixing of gas, slag and metal droplets takes place in and around the periphery of the gas jet. It has been confirmed at Hoogovens that the composition of the waste gases corresponds approximately to that in equilibrium with the activity of FeO around the metal droplet. The oxygen absorbed by the droplets in the form of an FeO layer is gradually transferred to bulk slag; the FeO layer on the metal may be ripped off either due to gas evolution (carbon monoxide gas formed by the reaction of carbon in the droplet reacts with the outer FeO layer) or to turbulent mixing conditions induced by the jet. Partially decarburized droplets fall back to the metal and their place is taken by new ones generated in the jet impact zone. A constant stream of droplets (iron conversion) is thus maintained throughout the blowing time.

Towards the end of the blow the FeO formed on the surface of the metal droplets begins to accumulate in the slag phase because of the lack of carbon in the ejected metal droplet and the slower rate of oxidation of carbon (or the reduction of FeO) at the slag–metal interface. On the basis of laboratory experiments it has been found that the reaction between a metal droplet and slag slows down appreciably as soon as the average carbon content of the droplet falls below 2 wt % [23].

The rate of entrainment of gas, emulsion, foam, droplet or slag into the oxidizing jet is a function of jet characteristics. The entrainment rate (or mixing rate), expressed as V_{slag} (tons/min), can be calculated from Equation (5.20) as

$$V_{slag} = (R_g - 1)F_{O_2}\rho_{O_2} \tag{5.38}$$

Since the metal droplets are heavier than slag they reduce the turbulence of slag and thereby decrease the transfer rate of FeO from the droplet to the bulk slag. The feed rate of droplets to the slag–droplet–gas mixture is the same as iron conversion (Mg in Equation (5.32)). The core of the droplets contains the reducing agents like carbon and silicon. An increase in iron conversion will lower the FeO content of the slag by promoting droplet–slag reaction. The dimensionless ratio of entrainment rate (V_{slag}) to feed rate (Mg) can be defined as the turbulence index of slag (TIS):

$$TIS = \frac{\text{entrainment rate of slag–droplet mixture into the jet (tons/min)}}{\text{rate of supply of material into slag for oxidation (tons/min)}}$$

$$= \frac{V_{slag}}{Mg} \text{ (dimensionless number)} \tag{5.39}$$

As a first approximation, the mass transfer coefficient in slag, k_{sl}, for the oxidation reactions due to the impinging gas jet and subsequent entrainment into the jet can be assumed to be proportional to the turbulence index of slag (TIS):

$$k_{sl} \propto (TIS)^{n_1} \text{ (m/s)} \tag{5.40}$$

The value of the exponent n_1 (ranging from 0.75 to 1.0) is found to depend upon the fraction of gas in the slag (gas hold-up or foaming index), the amount of undissolved lime and the mass of the slag; for a foamy slag n_1 is approximately 0.75. The proportionality constant in Equation (5.40) will depend upon the physical and geometrical parameters of the system including the volume ratio of metal and slag.

The total reaction surface (TRS) of ejected metal droplets decides the extent of slag–droplet reaction. Hence a decrease in TRS (for example at lance heights greater than 2.05 m in Figure 5.8), will lead to an increase in the FeO content of the slag due to reduced reactions between carbon in droplet in FeO in slag; this is observed in actual practice as well.

The volumetric mass transfer coefficient, $\mathcal{K}'_{sl}$, in slag for the oxidation of iron due to an impinging gas jet and the entrainment effect can be written as

$$\mathcal{K}'_{sl} = k_{sl} SRS \, Mg\tau_b \, (m^3/s) \tag{5.41}$$

The typical reaction time τ_b is given by

$$\tau_b = \frac{50 \times \text{bath weight}}{F_{O_2}} \, (\text{min})$$

Similar to the slag phase, the mixing rate of metal, V_{met}, can be defined as

$$V_{met} = \frac{\text{bath weight}}{\tau} \, (\text{ton/min}) \tag{5.42}$$

where τ is the mixing time calculated by taking into account the energy received by the bath from the top jet as well as from the bottom-stirring gas [24]. Under a steady-state process (i.e. in the absence of slopping) the feed rate of metal droplets to the bulk metal is nearly the same as iron conversion (Mg); the total amounts of iron, silicon, manganese, carbon and phosphorus which are oxidized per minute can be related to the molar flow of oxygen per minute. The turbulence index of metal (TIM) can be defined as

$$TIM = \frac{\text{recirculation stream along the reaction zone (tons/min)}}{\text{rate of supply of reacting material (tons/min)}}$$

$$= \frac{V_{met}}{\phi F_{O_2} \rho_{O_2}} \, (\text{dimensionless number}) \tag{5.43}$$

where ρ_{O_2} is the density of oxygen (tons/m^3) and ϕ is the ratio factor depending upon the blowing period.

As a first approximation, the overall turbulent mass transfer coefficient in the metal phase, k_{me}, for the reactions at the bulk metal–slag interface can be assumed to be proportional to the turbulence index of metal, TIM:

$$k_{me} \propto (TIM)^{n_2} \, (m/s) \tag{5.44}$$

The value of n_2 will depend upon the energy input to the bath, including the design of the stirring gas system and the amount of undissolved scrap present in the liquid metal. The constant of proportionality, similar to Equation (5.40), will depend upon physical and geometrical parameters of the system.

The net reaction area between slag and bulk metal (SRA) is obtained by subtracting the jet impact area ($\pi n I_A^2$) from the nominal area:

$$SRA = \pi(D_{con}^2 - n I_A^2)(m^2) \tag{5.45}$$

where D_{con} is both diameter, n is the number of nozzles and I_A is the impact diameter of each jet. The volumetric mass transfer coefficient for the reactions at bulk metal–slag interface ($\mathscr{K}_{me}'$) will be given by

$$\mathscr{K}_{me}' = k_{me}SRA(m^3/s) \tag{5.46}$$

The dimensionless ratio of $\mathscr{K}_{sl}'$ to $\mathscr{K}_{me}'$ is defined in the present work as the 'droplet ejection and oxidation' number or DEO number:

$$DEO\ number = \mathscr{K}_{sl}'/\mathscr{K}_{me}' \tag{5.47}$$

Extensive slag formation studies have been carried out at Hoogovens. The slag samples taken during the first phase of the blow (0–25% of blow time) indicate that a silica-rich slag forms on top of the metal. If the ejected metal droplets are caught up in the slag in the first few minutes of the blow then the apparent slag viscosity increases, lime dissolution slows down and a dry, hard slag begins to form. The droplets generated in the jet impact region are then no longer arrested by the slag and instead, as visually observed on the shop floor, they are free to travel towards the lance, converter mouth, skirt and gas recovery hood thereby causing skulling, increased lining wear, etc. The lance height is kept at maximum (subject to jet coalescence) during the initial stages of the blow to promote slag formation.

An average iron content of 8–12% in the bulk slag during the main blow period (25–75% of blow time) helps to maintain the slag fluidity thereby enhancing lime dissolution. The risk of forming solid, dry slags increases with low-FeO slags. The FeO content of the slag can be increased by raising the lance height, provided there is no risk of slopping.

Towards the end of the blow (75–100% of blow time) a deeper jet penetration helps in faster decarburization and in keeping the FeO content of the slag at a lower level (due to the increased recirculation of metal and increased iron conversion). Slag–metal equilibrium is slowly approached only towards the end of the blow. An increase in the lance height leads to an increase in the FeO content of the slag.

Each steel plant, based on practical experience and economical considerations, follows a set of specific blowing regimes. New blowing regimes and nozzle designs have to be tried out to arrive at the best combination of lance nozzle design and the blowing regime. The value of DEO increases with lance height (Figure 5.11, for lance head number 37 in plant 1). The pattern of variation of DEO for various blowing regimes (shown in Figure 5.7) and the corresponding nozzle designs listed in Table 5.1 is shown in Figure 5.12; since the values of n_1 and n_2 for different plants are not known, it has been assumed for the sake of simplicity that $n_1 = 0.75$, $n_2 = 1$ and $\phi = 1$. The pattern of variation of DEO is similar in all cases. It has been practically tested at Hoogovens that, for a given nozzle design and hot metal composition, the wear rate of the refractory lining, the wear pattern of the refractory lining, the

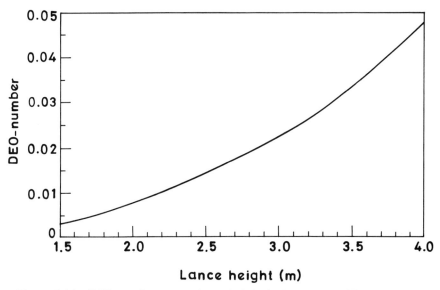

Figure 5.11   DEO number versus lance height for lance head 37 in Table 5.1.

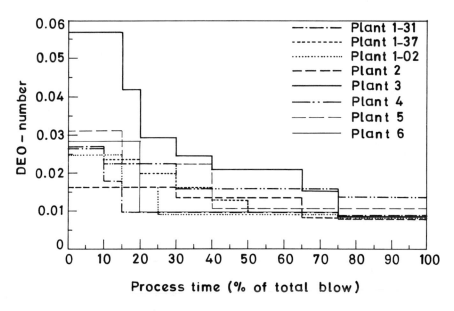

Figure 5.12   DEO number versus blowing time at various steel plants; see footnote of Table 5.1 for details of plants.

behaviour of slag formation and the turndown analysis can be reproduced when the blowing regime is adjusted to follow a specific pattern of variation of DEO number during a blow. The blowing regime for a new lance head design can thus be determined by calculating the lance height for a given DEO number.

Figure 5.12 includes both 100 and 300 ton converters. For a given nozzle design, the lance height can, in principle, be chosen so that the value of DEO is the same (hence the behaviour of slag formation is more or less similar). Care should, however, be taken to avoid jet coalescence at lance heights greater than 3 m. At low lance heights, $\Delta x/x_{bath} > 0.1$ should be followed to avoid deep jet penetration; Δx is depth of penetration and x_{bath} is bath depth.

Typical values of DEO during 0–25% of the blow range between 0.025 and 0.03, subject to jet coalescence; thus only in the case of specially designed nozzles can DEO be raised to higher values. In the main blow period (25–75% of blow time) DEO should be reduced in small steps to approximately 0.01–0.015. In the end blow period (75–100% of blow time), depending upon end point temperature, carbon and phosphorus, DEO should preferably be less than 0.01. If bottom stirring is employed, the effective lance height should be used to calculate DEO. Experimental measurements at Hoogovens with the help of a special sublance (to measure bath level as well as to collect slag and metal samples during the blow) have shown that the bath level rises (Figure 5.13) by approximately 7–10 cm during the peak period of bottom stirring.

The DEO number can also be used as a parameter to predict the actual FeO content of slag provided the oxygen potentials at various interfaces are known. In

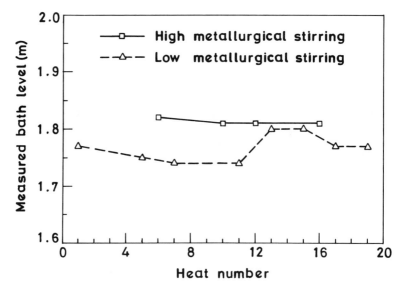

Figure 5.13 Effect of bottom-stirring gas on rise of bath level in 300 ton combined-blown converter at Hoogovens IJmuiden.

converter control slag–droplet model developed at Hoogovens, multicomponent mixed transport control theory [25, 26] is used to predict the oxygen potential at various interfaces.

5.5 Foaming of slags

The foaming of steelmaking slags, as in the formation of foam in an aerated soap solution, occurs as a result of the retention (or capture) of rising gas bubbles in the slag. Various studies on foams have been briefly reviewed by Ito and Fruehan [27].

In top-blown oxygen steelmaking converters the CO/CO_2 gases are formed as a result of the oxidation of carbon mainly in the jet impact zone and at the droplet–slag and slag–metal interfaces. There is a continuous stream of droplets falling through the slag. The foam formed is actually a heterogeneous mixture of droplets, gas bubbles, undissolved flux particles (of lime, dolomite, fluorspar and iron ore) and liquid slag. The intensity of foaming in top-blown converters is sometimes so high that the foam comes out through the mouth of the converter. This phenomenon is technically termed 'slopping'. The control of foaming and slopping is necessary for a steady operation.

The intensity of foaming in bottom-blown converters is much less than in top-blown converters. A thick slag–metal–gas emulsion, rather than a foam, forms at the slag–metal interface; thin films of metal are carried into the slag phase by rising gas bubbles; the thin films disperse in the slag phase as ultra-fine particles; rising gas bubbles are captured in the slag; and the three-phase mixture of gas, metal and slag becomes an emulsion. The emulsion provides a very large area for the slag–metal–gas reaction and contact time is also increased. The bottom-blowing process, therefore, approaches equilibrium faster than the top-blowing process.

The foaming of slag in combined-blown converters is a net result of the foaming phenomenon in top and bottom blowing processes, as described above.

The slag foaminess can be expressed [27] as the average travelling time of the gas bubble through the foam and is mathematically expressed as 'foaming index' Σ (in seconds):

$$\Sigma = \Delta h / \Delta v_g^s \tag{5.48}$$

where h is foam height and v_g^s is superficial gas velocity. Laboratory experiments [27] have demonstrated that the principal parameters affecting the foaming behaviour of slags are: viscosity (μ, Pas), density (ρ, kg/m^3) and surface tension (σ, N/m). The empirical equation obtained from dimensional analysis is

$$\Sigma = 5.2 \times 10^2 \frac{\mu}{(\sigma\rho)^{1/2}} \tag{5.49}$$

Evidently, the foaming index is affected more by viscosity. It is reported that the foaming index for $CaO-SiO_2-20\%$ FeO slags (in the absence of metal droplets) may range from 0.6 to 1.3 [27].

Foam is stabilized when second-phase particles, like finely dispersed metal droplets, undissolved lime, dolomite, fluorspar, iron ore, precipitated dicalcium silicate, etc., are present in the slag. If ε is the volumetric fraction of second-phase particles, then the viscosity of slag may be estimated from [28]

$$\mu = \mu_f(1 + 5.5\varepsilon) \tag{5.50}$$

or alternatively from [29]

$$\mu = \mu_f(1 - \varepsilon)^{-2.5} \tag{5.51}$$

where μ_f is the viscosity without the second-phase particles.

On integrating Equation (5.48)

$$h = \Sigma(v_g^s - v_g^0) + h^0 \tag{5.52}$$

where v_g^0 is superficial gas velocity and h^0 is foam height when foaming begins. The value of foam height can thus be calculated.

It is difficult to quantify the values of surface tension (σ), viscosity (μ) and density (ρ) for an actual steelmaking slag which is a heterogeneous mixture of slag, metal droplets and undissolved flux particles. In addition to this, the slag phase is continuously recirculated by the impinging gas jet. With some approximations, however, Equation (5.52) has been adapted at Hoogovens IJmuiden to predict foam height as a function of blowing time in 300 ton converters; the main parameters responsible for foaming are estimated as follows:

- The amount of undissolved lime and precipitated dicalcium silicate are calculated by the dynamic slag formation model; this model incorporates the Kapoor–Frohberg model described in section 2.4.3.
- The size distribution and the amount of ejected metal droplets are calculated from the procedure described in section 5.4.

With the help of the above two models, approximate values of ε and μ are calculated. Surface tension values are taken from published sources. The value of v_g^s is calculated from the decarburization model (see Chapter 6) and the bottom gas (argon) flow rate. On substituting all the parameters the foam height can be calculated.

According to laboratory experiments [27], the calculated value of Σ for $CaO-FeO-SiO_2$ slags first decreases with basicity and reaches a minimum around $CaO/SiO_2 = 1.2$. With a further increase in basicity Σ attains a maximum value around $CaO/SiO_2 = 1.8$.

In a combined-blown 300 ton converter at Hoogovens IJmuiden, foaming steadily increases after 25% of the blowing time and attains a maximum value around the middle of the blow (i.e. after 50–60% of the blowing time). The foam begins to die after 75% of the blowing time; during 25–75% of the blowing time, the basicity of the slag steadily increases from 1.2 to approximately 2 while the FeO content of the slag is almost constant (between 5 and 8 wt %).

Lance height has a great effect on foaming index because it affects the rate of generation of metal droplets as well as the recirculation of slag and gas within the vessel. Lance height adjustment can be used as a tool to control the foaming of slag; if lance height is decreased then foaming decreases and therefore, during slopping, the lance is lowered.

Larger additions of CaF_2 decrease foam stability by reducing the viscosity (μ) of the slag. The addition of iron ore can increase the foaming of slag.

The effect of the addition of MgO on foaming is not clearly understood. On the one hand, MgO decreases the viscosity of the slag due to a decrease in the melting point. On the other hand MgO retards the dissolution rate of lime and thereby increases the proportion of undissolved lime (i.e. the proportion of second-phase particles).

References

1. Van Hoorn, A. I., van Konijnenburg, J. T. and Kreijger, P. J. (1976) 'Evolution of slag composition and weight during the blow' In *The Role of Slag in Basic Oxygen Steelmaking Processes* (ed. W.-K. Lu), *Symp. Proc. at McMaster University, Hamilton, Canada, 1976*, pp. 2.1–2.22.
2. Bardenheuer, F., vom Ende, H. and Speith, K. G. (1970) 'Contribution to the metallurgy of the LD process', *Blast Furnace and Steel Plant*, **58**, 401–6.
3. Obst, K. H., Schürmann, E., Münchberg, W., Mahn, G. and Nolle, D. (1980) 'Auflösungsuerhalten und Sattigungsgehalte des Magnesium oxides in Anfangsschlacken des LD-Verfahrens' *Arch. Eisenhüttenwes.*, **51**, 407–12.
4. Williams, P., Sunderland, M. and Briggs, G. (1982) 'Interactions between dolomitic lime and iron silicate melts', *Ironmaking and Steelmaking*, **9**, 150–62.
5. Matsushima, M., Yadoomaru, S., Mori, K. and Kawai, Y. (1977) 'A fundamental study on the dissolution rate of solid lime into liquid slag', *Trans. Iron Steel Inst. Jpn*, **17**, 442–9.
6. Umakoshi, M., Mori, K. and Kawai, Y. (1984) 'Dissolution rate of burnt dolomite in molten Fe_tO–CaO–SiO_2 slags', *Trans. Iron Steel Inst. Jpn*, **24**, 532–9.
7. Oeters, F. and Scheel, R. (1974) 'Untersuchungen zur Kalkauflösung in CaO-FeO-SiO_2-Schlacken' *Arch. Eisenhüttenwes.*, **45**, 575–80.
8. Boom, R., Deo, B., van der Knoop, W., Mensonides, F. and van Unen, G. (1988) 'The application of models for slag formation in basic oxygen steelmaking', *3rd Int. Conf. on Metallurgical Slags and Fluxes: Symp. Proc, Institute of Metals, Glasgow, 27–29 June, 1988*, pp. 273–6..
9. Koria, S. C. (1988) 'Dynamic variations of lance distance in impinging jet steelmaking practice', *Steel Res.*, **59**, 257–62.
10. Young, P. A. (1967) 'The calculability of iron and steelmaking processes', *J. Aust. Inst. Met.*, **12**, 271–91.
11. Inada, S. and Watanabe, T. (1977) 'Influence of pressure in surrounding atmosphere on the cavity formation by gas jet', *Trans. Iron Steel Inst. Jpn*, **17**, 59–66.

12. Chatterjee, A. (1972) 'On some aspects of supersonic jets of interest in LD steelmaking, Part 2', *Iron and Steel*, December, 627–34.

13. Koria, S. C. and Lange, K. W. (1984) 'A new approach to investigate the drop size distribution in basic oxygen steelmaking', *Metall. Trans.*, **15B**, 109–16.

14. Koria, S. C. and Lange, K. W. (1986) 'Estimation of drop sizes in impinging jet steelmaking processes', *Ironmaking and Steelmaking*, **13**, 236–40.

15. Schoop, J., Resch, W. and Mahn, G. (1978) 'Reactions occurring during the oxygen top-blown process and calculation of metallurgical control parameters', *Ironmaking and Steelmaking*, **2**, 72–9.

16. He, Q. L. and Standish, N. (1990) 'A model study of droplet generation in the BOF steelmaking', *Iron Steel Inst. Jpn Int.*, **30**, 305–9.

17. Sharma, S. K., Hlinka, J. W. and Kern, D. W. (1977) 'The bath circulation, jet penetration and high-temperature reaction zone in BOF steelmaking', *Ironmaking and Steelmaking*, July, 7–18.

18. Koria, S. C. and Lange, K. W. (1987) 'Penetrability of impinging gas jets in molten steel bath', *Steel Res.*, **58**, 421–6.

19. Chatterjee, A. (1973) 'On some aspects of supersonic jets of interest in LD steelmaking, Part 2', *Iron and Steel*, February, 38–40.

20. Bird, R. Byron., Steward, E. W. and Lightfoot, E. N. (1960) *Transport Phenomena*, Wiley: New York, pp. 157–67.

21. Molloy, N. A. (1970) 'Impinging jet flow in a two-phase system: The basic flow pattern', *J. Iron Steel Inst.*, October, 943–50.

22. Lee, C. K., Neilson, J. H. and Gilchrist, A. (1977) 'Effect of nozzle angle on the performance of multi-nozzle lances in steel converters', *Ironmaking and Steelmaking*, **4**, 329–37.

23. Sawada, Y., Krishna Murthy, G. G. and Elliott, J. F. (1992) 'Reduction of FeO dissolved in $CaO–SiO_2–Al_2O_3$ slags by iron carbon droplets', *EPD Congr.*, *The Minerals, Metals and Materials Society* (ed. J. P. Hager), pp. 915–29.

24. Lange, K. W. (1988) 'Thermodynamic and kinetic aspects of secondary steelmaking processes', *Int. Mater. Rev.* **33**, 53–89.

25. Robertson, D. G. C., Deo, B. and Ohguchi, S. (1984) 'Multicomponent mixed-transport-control theory for kinetics of coupled slag/metal and slag/metal/gas reactions: application to desulphurization of iron', *Ironmaking and Steelmaking*, **11**, 41–55.

26. Deo, B., Ranjan, P. and Kumar, A. (1987) 'Mathematical model for computer simulation and control of steelmaking', *Steel Res.*, **58**, 427–31.

27. Ito, K. and Fruehan, R. J. (1988) 'Study on the foaming of $CaO–SiO_2–FeO$ slags: Part 1. Foaming parameters and experimental results', *Metall. Trans.*, **20B**, 509–14; 'Part 2. Dimensional analysis and foaming in iron and steelmaking processes', 515–21.

28. Birkman, H. C. (1952) 'The viscosity of concentrated solutions and suspensions', *J. Chem. Phys.*, **20**, 571.

29. Happel, J. (1958) 'Viscous flow in multiparticle systems', *Am. Inst. Chem. Eng. J.* **4**, 197–201.

6 *Primary steelmaking metallurgy: refining reactions*

6.1 Reacting phases and interfaces in oxygen steelmaking

The topic of refining reactions in oxygen steelmaking continues to remain as interesting and intriguing as ever. A schematic diagram of the physical state of the LD vessel [1] towards the middle of a refining period is shown in Figure 6.1. Many reacting sites and reaction mechanisms have been proposed by different workers. As a result, a large amount of published information on both the theory and practice of oxygen steelmaking is available in the literature. Only selected reviews and research papers relating to fundamentals will be quoted in this chapter for the sake of a clear understanding of the concepts involved.

Important chemical reactions which determine the partitioning of oxygen in slag, metal and gas phases [2] are given in Figure 6.2 and Table 6.1. It is clear that the refining reactions in oxygen steelmaking can take place at any one of the following sites [1] in a simultaneous or sequential manner:

- within the jet;
- directly under the jet;
- in the cavity emulsion or cavity four-phase (liquid slag, metal, gas, undissolved lime or solid slag) dispersion;
- in the slag–metal emulsion in the bath;
- in the slag foam between droplets and the oxidizing slag;
- along the effluent gas trajectory;
- in the atmosphere above the bath and slag;
- at the interfaces between the oxidizing slag (or gas) and iron, away from the cavity;
- on the refractory walls;
- at the interface of undissolved scrap and liquid metal;
- at the interface of undissolved lime (or solid slag) and metal.

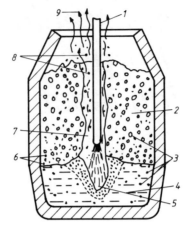

Figure 6.1   Physical state of LD vessel during the middle (50% of blow time) of a heat: 1, lance; 2, gas–slag–metal emulsion; 3, carbon monoxide bubbles; 4, metal; 5, entrained material in oxygen jet; 6, metal droplets; 7, carbon monoxide gas formed in jet impact zone; 8, spray of metal entrained in outgoing carbon monoxide gas stream; 9, brown smoke. (*Source*: Lange, K. W. (1981) *Physical and Chemical Influences during Mass Transfer in Oxygen Steelmaking Process*, West Deutscher Verlag: Lingerich.)

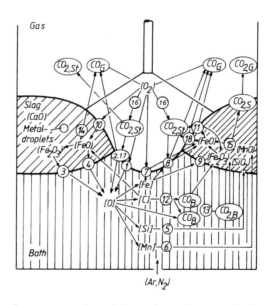

Figure 6.2   Schematic representation of sites of reaction and distribution of oxygen in various reactions in a top-blown converter. (*Source*: Lange, K. W. (1981) *Physical and Chemical Influences during Mass Transfer in Oxygen Steelmaking Process*, West Deutscher Verlag: Lingerich.)

Table 6.1 Fe–C–O reaction at the interfaces of slag, metal and gas.

		Equation
[O] pickup by bath:		
(O_2) in hot spot	$\frac{1}{2}(O_2) = [O]$	(1)
(CO_2) in hot spot	$(CO_2)_{st} = [O] + (CO)_g$	(2)
(Fe_2O_3) of slag	$(Fe_2O_3) = [O] + 2(FeO)$	(3)
(FeO) of slag	$(FeO) = [O] + [Fe]$	(4)
Removal into slag phase:		
[Si] through [O] in bath	$[Si] + 2[O] = (SiO_2)$	(5)
[Mn] through [O] in bath	$[Mn] + [O] = (MnO)$	(6)
[Fe] through (O_2) in hot spot	$[Fe] + \frac{1}{2}(O_2) = (FeO)$	(7)
[Fe] through (CO_2) in hot spot	$[Fe] + (CO_2)_{st} = (FeO) + (CO)_g$	(8)
[Fe] through (Fe_2O_3) of slag	$[Fe] + (Fe_2O_3) = 3(FeO)$	(9)
(O) pickup by slag:		
(O_2) in hot spot	$2(FeO) + \frac{1}{2}(O_2) = (Fe_2O_3)$	(10)
(CO_2) in hot spot	$2(FeO) + (CO_2)_{st} = (Fe_2O_3) + (CO)_g$	(11)
Oxidation of [C] to (CO)		
and (CO_2):		
in bath	$[C] + [O] = (CO)_{st}$	(12)
in bath	$(CO)_{st} + [O] = (CO_2)_{st}$	(13)
in slag (metal droplets)	$[C] + (FeO) = (CO)_{st} + [Fe]$	(14)
in slag	$(CO)_{st} + (Fe_2O_3) = (CO_2)_{st} + 2(FeO)$	(15)
Postcombustion of (CO) to (CO_2):		
through O_2 jet	$(CO) + \frac{1}{2}(O_2) = (CO_2)_{st}$	(16)
Decomposition of (CO_2) obtained		
from postcombustion:		
in hot spot	$(CO_2)_{st} = (CO) + [O]$	(17)
in slag	$(CO_2)_{sl} + 2(FeO) = (CO)_{sl} + (Fe_2O_3)$	(18)

Sites of various reactions: summary

(CO_2) formation	(13), (15), (16)
(CO_2) consumption	(8), (17), (18)
[O] pickup by bath	(1), (2), (3), (4)
[O] consumption	(5), (6), (12), (13)
(O) pickup by slag	(10), (11)
(O) given by slag	(3), (4), (14), (15)

Note: 'st' indicates steel, 'sl' indicates slag and 'g' indicates gas phase. Equation numbers refer to Figure 6.2

Source: Schürmann, E., Sperl, H., Hammer, R. and Ender, A. (1986) *Stahl und Eisen*, **106**, 1267–72.

The most interesting aspect of the oxygen steelmaking process is that the partitioning of oxygen among the above sites and also the kinetics of reactions are affected by the conditions of slag formation (i.e. solid slag, liquid slag, foaming slag, highly oxidizing slag, etc.) during the progress of a blow. The refining reactions considered in this chapter are oxidation of carbon, silicon, manganese, sulphur and phosphorus.

6.2 Carbon–oxygen reaction

The thermodynamics of carbon and oxygen dissolution in molten iron has already been discussed in Chapter 2. Various possible reactions leading to the removal of carbon from metal (with reference to Figure 6.2 and Table 6.1) are as follows:

- In the jet impact region by direct oxidation of carbon at the gas–metal interface:

$$[C] + \tfrac{1}{2}(O_2)_g = (CO)_g \qquad (6.1)$$

$$[C] + (CO_2)_g = 2(CO)_g \qquad (6.2)$$

- In the emulsion, foam or at slag–metal–gas interfaces through a series of reaction steps:

$$\{Fe\} + \tfrac{1}{2}(O_2)_g = (FeO) \qquad (6.3)$$

$$(CO)_g + (FeO) = (CO_2)_g + \{Fe\} \qquad (6.4)$$

$$\underline{(CO_2)_g + [C] = 2(CO)_g} \qquad (6.5)$$

$$(FeO) + [C] = \{Fe\} + (CO)_g \qquad (6.6)$$

- Within the metal at the bubble–metal interface:

$$[C] + [O] = (CO)_g \qquad (6.7)$$

The bubble may contain $(CO + CO_2)$ or $(CO + CO_2 + \text{inert gas})$. The dissolved oxygen, $[O]$, may enter the metal through the following routes:

- direct absorption in the jet impact region,

$$(O_2)_g = 2[O] \qquad (6.8)$$

$$(CO_2)_g = [O] + (CO)_g \qquad (6.9)$$

- via the FeO present in the slag phase or formed in the jet impact region,

$$[Fe] + \tfrac{1}{2}(O_2)_g = (FeO) \quad \text{(in the jet impact zone)} \qquad (6.10)$$

$$(FeO) = [Fe] + [O] \quad \text{(FeO is present in the slag)} \qquad (6.11)$$

The rates of the above reactions, Equations (6.1)–(6.11), may be controlled by chemical kinetics and/or mass transport steps. Thus, bulk mixing, surface renewal and interfacial factors (surface tension, chemical poisoning) can contribute to the overall kinetics. Some important practical observations on the carbon–oxygen reaction (as summarized below) are already well known and provide an idea of the predominant mechanism(s) of carbon oxidation which could be operating at different stages of the blow:

1. A typical profile of the carbon oxidation rate (dC/dt) versus dissolved carbon and oxygen contents in metal is shown in Figure 6.3(a) [3]. The profile can

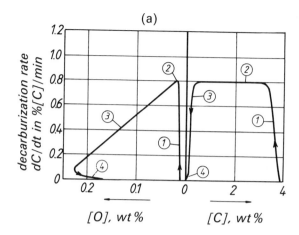

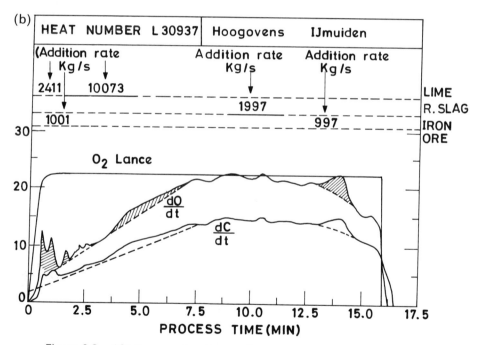

Figure 6.3 d*C*/d*t* curve at various carbon concentrations: (a) schematic representation; (b) actual curve for a 300 ton vessel as a function of blow time; the shaded area shows the extra oxygen liberated as a result of ore dissolution. (Figure 6.3(a) *Source*: Koch, K., Fix, W. and Valentin, P. (1976) *Arch. Eisenhüttenwes.*, **47**, 659–63.)

be divided into four distinct regimes. In the initial blow period (regime 1) the carbon oxidation rate increases linearly; in the main blow period (regime 2) it remains constant; in the end blow period (regime 3) it decreases linearly until it reaches a certain limit and then it drops almost asymptotically (regime 4) with time. The characteristic slopes and the durations of regimes 1–4 in Figure 6.3(a) have also been found to be dependent on parameters specific to the process, like converter size, oxygen flow rate, lance type, etc. The dC/dt and dO/dt curves observed in a 300 ton converter at Hoogovens IJmuiden are shown as a function of blowing time in Figure 6.3(b).

2. The carbon content of metal at the intersection of regimes 2 and 3 is generally referred to as the critical carbon content, C_{crit}. Figure 6.4 shows an increase of C_{crit} with oxygen flow rate [1]. The effect of lance type on C_{crit} is not yet clearly understood [1, 3].

3. In regime 1 the slope of dC/dt versus time depends (for a fixed lance height) mainly on the metal silicon content and oxygen flow rate because in the initial part of the blow most of the oxygen is consumed by silicon. Also, a steep rise in FeO content of slag takes place (Figure 5.2); this may be attributed to the difficulty of nucleation of carbon monoxide bubbles at the slag–metal interface in the presence of silicon in the metal. The change of lance height affects the oxygen utilization between iron, manganese, phosphorus and silicon and carbon (and hence dC/dt).

4. The peak value of dC/dt and its duration in the main blow period, regime 2, depends upon the oxygen flow rate [4], as shown schematically in Figure 6.5.

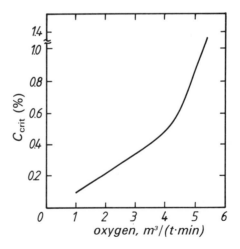

Figure 6.4   Critical carbon content C_{crit} (for the onset of regime 3 in Figure 6.3(a)) as a function of specific oxygen consumption. (*Source*: Koch, K., Fix, W. and Valentin, P. (1978) *Arch. Eisenhüttenwes.*, **49**, 231–4.)

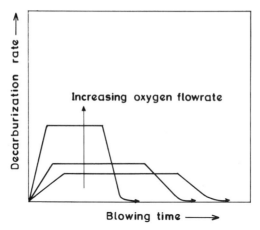

Figure 6.5 Schematic representation of dC/dt curves at increasing oxygen flow rates.

Despite the rise in dC/dt with increasing oxygen flow rates the pattern of evolution of percentage carbon dioxide in exhaust gas defined as [5]

$$\%CO_2' = \left(\frac{\%CO_2}{\%CO + \%CO_2} \right) \times 100$$

increases with lance height (Figure 6.6a); if the lance height is kept fixed then the pattern is unchanged implying that the mechanism of postcombustion reactions remains unaffected (Figure 6.6b). The type of scrap charged also affects the postcombustion. Light scrap melts faster than heavy scrap [5]. Thus at 60–70% of the blow, in Figure 6.6(c), most of the light scrap would be molten whereas the presence of heavy unmelted scrap retards the metal circulation velocities and thereby affects the depth of penetration of the jet. The profile of the dC/dt versus time curve in regime 2 still remains unaffected; only C_{crit} and postcombustion are affected by the type of scrap.

5. Koch *et al.* [4] successfully measured the hot spot (jet impact zone) temperature in 50 kg top-blowing experiments by employing pure iron–carbon alloys with no slag cover. The carbon and oxygen contents of the metal as well as the waste gas temperatures were measured and the results [6, 7] are shown in Figure 6.7. During the main blow period, the hot spot temperature was 2300 °C and the waste gas temperature was 2100 °C. The value of dC/dt was constant for a given flow rate of oxygen, but increased when oxygen flow rate was increased (Figure 6.5 and Figure 6.8). As soon as dC/dt dropped towards the end of the main blow, the hot spot temperature decreased and the oxygen content of the metal began to rise – signalling a sudden change in the mechanism of carbon oxidation and oxygen transfer to metal at C_{crit}; with increasing oxygen flow rate dC/dt dropped down faster (Figure 6.5) in regime 3.

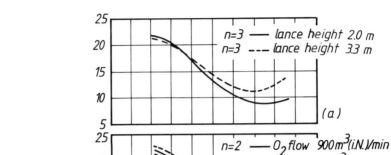
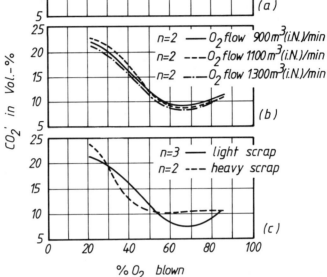

Figure 6.6 Carbon dioxide content in waste gas as a function of blow time: (a) effect of lance height variation; (b) effect of oxygen flow rate variation; (c) effect of type of scrap. (*Source*: Lange, K. W. (1981) *Physical and Chemical Influences during Mass Transfer in Oxygen Steelmaking Process*, West Deutscher Verlag: Lingerich.)

6. Towards the end of the blow in regime 4, the dissolved oxygen in the metal began to decrease (Figure 6.7), but always remained above the equilibrium value. The equilibrium oxygen content is lower if the bath is stirred with argon; the carbon and oxygen product in the case of argon stirring is 20×10^{-4} whereas in the absence of argon stirring [8] it rises to 35×10^{-4} (Figure 6.9).

7. The maximum temperatures developed in the hot spot region in an actual converter have not been measured because of the presence of the slag phase; slag interacts with the jet and also removes heat from the hot spot. It is found that the slag is at a higher temperature (at least $50\,°C$) than the metal throughout the blow.

8. The product of carbon oxidation in the hot spot region is carbon monoxide only. The metal droplets created in the impingement zone and their further disintegration into smaller droplets due to evolution of carbon monoxide

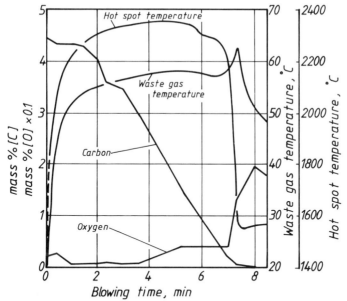

Figure 6.7 Hot spot temperature as a function of blow time; note the decrease in hot spot temperature after 7 minutes (*Source*: Koch, K., Fix, W. and Valentin, P. (1976) *Arch. Eisenhüttenwes.*, **47**, 102–14; **49**, (1978) 163–6.)

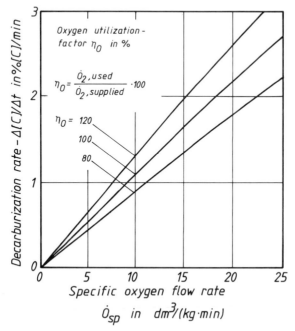

Figure 6.8 Increase in dC/dt due to increase in specific oxygen flow rate. (*Source*: Koch, K., Fix, W. and Valentin, P. (1976) *Arch. Eisenhüttenwes.*, **47**, 102–14; **49**, (1978) 163–6.)

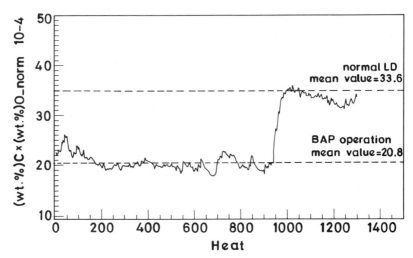

Figure 6.9 Oxygen activity as a function of carbon content at tapping in top- and combined-blown oxygen steelmaking processes.

during their passage through the slag have been shown to create a large slag–metal interfacial area [9]. In spite of this, however, a significant contribution of the slag phase to the overall decarburization is doubtful; the following practical observations have been made at Hoogovens [8]:

(a) dC/dt in the main blow for a given flow rate of oxygen does not change with the nature of slag, i.e. solid (dry) or liquid or foamy slags.

(b) Even the total amount of FeO present in the slag during the main blow is such that it could hardly contribute to more than 20% of decarburization unless it is postulated that FeO is being constantly supplied from the hot spot region and then reduced at the slag–metal interface, away from the hot spot. This is in contradiction with observation (a). In large converters the FeO formed during the first 2–3 minutes is rapidly reduced to a level of 5–7% (Figure 5.2). Further, experiments [9] have demonstrated that the rate of decarburization of metal droplets by slag decreases rapidly as the FeO content of the slag is lowered. This is attributed to the problem of carbon monoxide nucleation at the slag–metal interface.

The conclusions which emerge from the above practical observations can be summarized as follows:

1. Main blow period (Regime 2)

(a)　dC/dt is independent of carbon level (Figure 6.3);
(b)　dC/dt increases with oxygen flow rate (Figures 6.5 and 6.8);
(c)　dissolved oxygen in metal is extremely small (Figure 6.7);

(d)   the contribution of the slag phase to decarburization is less than 20% and most of the decarburization takes place in the hot spot region and within the droplets;

(e) C_{crit} increases with oxygen flow rate (Figure 6.4);

(f)   the effect of type of lance nozzle on dC/dt or C_{crit} is not clear;

(g)   the percentage of carbon dioxide in the waste gases depends essentially on the lance height and the type of scrap charged (Figure 6.5).

The above aspects lead us to think that in the main blow period, gas phase mass transfer and/or chemical reaction can be a rate-controlling step for the oxidation of carbon (at high carbon levels above 0.5 wt %).

2. End blow period (Regime 3)

(a) dC/dt is proportional to carbon content (Figure 6.5);

(b) dC/dt versus time drops faster with an increase in oxygen flow rate (Figure 6.5);

(c)   dissolved oxygen in metal increases (Figure 6.7), but if argon stirring or bottom blowing is used the oxygen concentration does not rise as much;

(d)   foam or emulsion dies down and FeO in the slag increases; owing to low carbon in the metal droplets, the contribution of the slag phase to decarburization should be negligible.

All these observations indicate that in regime 3 the mass transfer of species in gas and/or metal can be rate controlling.

3. Regime 4   When carbon has decreased to a low value, say below 0.5%, metal phase mass transfer control is important as the decarburization reaction is enhanced by argon bubbling. Decarburization at low carbon contents is discussed in detail in Chapter 7.

6.2.1   *Oxidation of dissolved carbon by gaseous oxygen and CO/CO_2 gas mixture*

The rate equations for the oxidation of dissolved carbon (in high-carbon melts) by gaseous oxygen [10–15] are summarized in Appendix C.

    The periphery of the oxygen jet in an oxygen steelmaking converter is contaminated as a result of the entrainment of gases from the surrounding atmosphere (mostly carbon monoxide gas). The following mechanisms have been proposed for specific cases of decarburization by CO/CO_2 gas mixtures:

1. At high gas flow rates of carbon monoxide and carbon dioxide and at a high carbon content in the metal (above C_{crit} but in the absence of sulphur), the dissociative absorption of carbon dioxide at the interface is rate controlling [12].

2. In the presence of sulphur the mixed control mechanism involving gas phase mass transfer and the dissociative absorption of carbon dioxide at the interface becomes rate controlling [13, 16].
3. At low carbon content the reaction is controlled by mixed mass transport in the metal and gas; it is predominantly metal phase control to start with but shifts towards the end to mass transport control in the gas phase [17].
4. Experiments with $^{13}C^{18}O$ isotope exchange [18] have shown that the chemical reaction rate at the interface is extremely fast compared with the mass transport in the gas phase.

The CO/CO_2 gas can form a film around the metal droplets traversing the slag. Alternatively, the CO/CO_2 gas bubbles themselves may sit at or traverse the slag–metal interface. These cases are discussed below.

Gas film around metal droplets

The metal droplets are ejected as a result of impingement of the supersonic oxygen jet on the metal surface. The number, size and retention time of these droplets in the slag can vary over a large range depending upon the operating conditions (see Chapter 5). It has been observed that when a carbon-containing metal droplet comes into contact with slag containing FeO it is immediately enveloped by a CO/CO_2 gas film [9]. The reaction at the gas–metal interface is

$$(CO_2)_g + [C] = 2(CO)_g \tag{6.12}$$

whereas at the gas–slag interface the region is

$$(CO)_g + (FeO) = (CO_2)_g + \{Fe\} \quad \text{(same as Equation (6.4))} \tag{6.13}$$

The choice of a particular rate equation will depend on the conditions of the experiment. For example, an optical examination of the droplets entrained in slag shows the presence of a thin decarburized layer at the surface indicating that the outward diffusion of carbon is unable to cope with its oxidation rate. Thus, in this case, the outward mass transfer of carbon in the metal and inward transfer of oxygen can be rate limiting.

Gas film at the slag–metal interface

A kinetic study of the reduction of 2.5% FeO slags by carbon in iron [19] has confirmed that dissociative chemisorption of carbon dioxide at the gas–metal interface is rate controlling [12]. At higher FeO contents, however, which is usually the case in oxygen steelmaking, it has been suggested that the chemical reaction at the gas–slag interface could be rate controlling [19]. The rate of gas–slag reaction is found to be independent of FeO concentration in the range 67.7 to 48% FeO, but below 48% FeO the rate is dependent on FeO concentration and carbon monoxide pressure, i.e. mixed control (mass transfer in slag plus chemical reaction at interface) [20]. Slag viscosity and the activity of FeO have considerable influence on the gas–slag reaction (i.e. mass transfer in slag) [21, 22]. During steelmaking, the temperature increases gradually from 1350 to 1650 °C. At high temperatures (say 1600 °C) and low carbon contents (below C_{crit}) it is also likely that mixed mass transport in the slag, gas and metal is rate controlling. In developing an appropriate kinetic model for decarburization by the $CO–CO_2$ gas film at the slag–metal interface, a coupling mechanism for simultaneous reactions at both the interfaces (gas–metal and gas–slag) is required. One such approach of coupled reactions assuming mixed mass transport control is described in section 6.7.

The rate equations for the oxidation of dissolved carbon by a CO/CO_2 gas mixture [12, 13, 16–18] are summarized in Appendix D.

6.3 Oxidation of silicon

The thermodynamics of silicon dissolution in molten iron has been discussed in Chapter 2.

The rate of oxidation of silicon at the slag–metal interface is controlled by its mass transfer in the metal to the slag–metal interface where it reacts with FeO in the slag:

$$[Si] + 2(FeO) = (SiO_2) + 2\{Fe\} \tag{6.14}$$

A part of the silicon is also oxidized directly in the jet impact zone:

$$[Si] + (O_2)_g = (SiO_2) \tag{6.15}$$

The kinetic equations for metal phase control have been discussed in Chapter 3. Also, the kinetics of reaction (6.14) has been discussed in Example 3.4.

As long as silicon is present in the metal, say above 0.05 wt %, it suppresses the formation of carbon monoxide gas at the slag–metal interface (or at the metal droplet–slag interface) and this leads to a good slag–metal contact for reaction (6.14) to proceed at a fast rate (regime 1 in Figure 6.3a).

It is commonly assumed that in the steelmaking slags the initial product of the reaction of SiO_2 with lime is $2CaO \cdot SiO_2$:

$$(SiO_2) + 2(CaO) = (2CaO \cdot SiO_2) \tag{6.16}$$

The rate of oxidation of silicon is, however, independent of the quality of lime (or type, namely powdered or lumpy) or the path of slag formation.

6.4   Oxidation of manganese

Besides the direct oxidation in the jet impact zone, the oxidation of manganese can take place via two routes:

$$[Mn] + [O] = (MnO) \tag{6.17}$$

$$[Mn] + (FeO) = (MnO) + \{Fe\} \tag{6.18}$$

The first reaction, Equation (6.17), can take place only at the beginning of the blow when the manganese concentration is high and the metal temperature is low. During the major part of refining the oxidation of manganese proceeds via Equation (6.18). The equilibrium constant can be written as

$$K = \frac{a_{MnO}}{h_{Mn} a_{FeO}}$$

$$\lg K = (6440/T) - 2.95 \tag{6.19}$$

since

$$a_{FeO} = \frac{[O]}{[O]_{sat}} \quad \lg[O]_{sat} = -(6320/T) + 2.734 \tag{6.20}$$

$$a_{MnO} = K h_{Mn} \frac{[O]}{[O]_{sat}} \tag{6.21}$$

where $[O]_{sat}$ can be calculated from Equation (6.20). Thus, knowing a_{MnO} we can calculate the equilibrium manganese content of steel. The kinetic equations for manganese oxidation via Equation (6.18) have already been discussed in Example 3.3. Coupled slag–metal–gas reactions involving silicon, manganese, carbon and iron are discussed later in section 6.7.

Enhanced oxidation of manganese at the beginning of the blow for a top-blown process (Figure 5.1) could occur by oxidation of the manganese in the droplets through Equation (6.17). In the case of a bottom-blown process manganese oxidation occurs via gas–vapour reactions between oxygen and the vapours of components at tuyères, namely

$$Fe(V) + Mn(V) + O_2(g) = FeO, MnO(l) \tag{6.22}$$

$$Fe(V) + Mn(V) + SiO(V) + \tfrac{3}{2}O_2(g) = FeO, MnO, SiO_2(l) \tag{6.23}$$

where (V) indicates vapour phase and (l) indicates liquid phase.

If lime is injected along with oxygen through the tuyères at the bottom, the silicon monoxide reacts preferably with the lime. Partial manganese oxidation takes place only after the silicon has disappeared. During the last stages of the blowing, Equation (6.18) becomes preponderant due to the decrease of the CO/CO_2 gas film around the metal droplets; the reaction more or less approaches equilibrium. In general, a lower FeO content in the final slag results in higher manganese recovery and this is also one of the advantages claimed for the bottom/combined-blown processes.

The combined effect of the basicity of slag and FeO on manganese distribution [23] is shown in Figure 6.10. The dashed line represents equilibrium corresponding to Equation (6.18). The apparent equilibrium constant is given by

$$k_{\text{Mn–Fe}} = \frac{(\% \, \text{MnO})}{[\% \, \text{Mn}](\% \, \text{FeO})} \frac{(\% \, \text{CaO})}{(\% \, \text{SiO}_2)} = 6 \tag{6.24}$$

The experimental data corresponding to Equation (6.24) are found to be very closely scattered around the equilibrium line [23], and this implies that the manganese reaction approaches equilibrium in oxygen steelmaking. In bottom/combined-blown processes the concentration of MnO in slag is lower than in LD because in the former the MnO formed near the tuyère is reduced by carbon in the metal bath while the slag is rising to the surface.

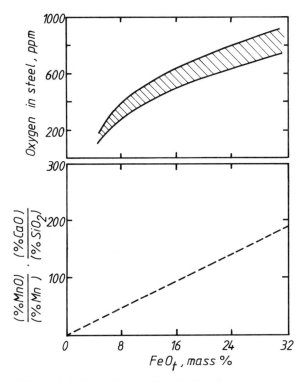

Figure 6.10   Effect of FeO content of slag on dissolved oxygen in steel and on manganese distribution. (*Source*: Turkdogan, E. T. (1984) *Trans. Iron Steel Inst. Jpn,* **24**, 591–611.)

6.5 Oxidation of phosphorus

The partitioning of phosphorus between the slag and metal is known to be very sensitive to process conditions and so far it has not been possible to build a kinetic model based on simple assumptions.

The dependence of phosphorus distribution on the FeO content of slag in LD and Q-BOP is shown in Figure 6.11. A parameter k_{ps} is defined [23] as

$$k_{ps} = \frac{(\% P_2 O_5)}{[\% P]_{metal}} (1 + (\% SiO_2)) = \phi((\% FeO), B) \tag{6.25}$$

where B is basicity; for $B > 2.5$, k_{ps} is found to be independent of B. Regression equations have also been proposed in terms of optical basicity [24].

The distribution of phosphorus is also found to be related to carbon in steel at the time of tapping; owing to lower carbon levels achieved in bottom-blown processes, the phosphorus distribution is expected to be better than in LD. In general, high basicity and the low temperature of slag (irrespective of the FeO content) favour dephosphorization. The fluxing agents, namely ilmenite or fluorspar,

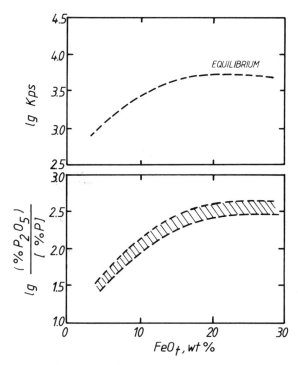

Figure 6.11 Effect of FeO content of slag on phosphorus distribution and lg k_{ps} value. (*Source*: Turkdogan, E. T. (1984) *Trans. Iron Steel Inst. Jpn*, **24**, 591–611.)

are added to decrease slag viscosity for better dephosphorization. The reaction rate can be further enhanced if lime is injected in the form of a powder through tuyères at the bottom together with oxygen.

Practical experience at Hoogovens has shown that slag formation is the key to phosphorus control in combined-blown converters. Phosphorus can be captured in the slag during the very early stages of the blow through judicious adjustment of the blowing practices through DEO number (Chapter 5). The dicalcium silicate (C_2S) formed in the early stages of the blow has solid solubility [25] for P_2O_5 (up to 2.6%). Blowing can be carried out such that the lighter C_2S crystals float up in the FeO-rich slag formed at the beginning of the blow (Figure 5.2). The C_2S crystals do not melt but remain there until, at three-quarters of the blow, FeO in the slag starts increasing again and a fluid slag forms. Thus, in the middle part of the blow, when the conditions of phosphorus reversal exist due to the reduction of FeO and P_2O_5, phosphorus remains locked in solid slag particles of C_2S. This is easily achieved when low-silicon hot metal (less than 0.4% silicon) is used in large-capacity converters; at low silicon content, solid slag forms early in the blow (Figure 5.5) and the slag volume is small. With such a non-equilibrium approach the quantity of lime needed to attain the required degree of dephosphorization is also reduced. It was further found at Hoogovens that low phosphorus levels (below 0.010 wt %) may not be achieved with dolomitic slags or with high-silicon hot metal because the mechanism of lime dissolution changes and slags with lower melting point and viscosity are formed.

6.6 Sulphur removal

Sulphur transfer takes place through the following reactions:

$$[S] + (O_2)_g = (SO_2)_g \tag{6.26}$$

(direct oxidation at the gas–metal interface in jet impact zone)

$$[S] + (O^{2-}) = (S^{2-}) + [O] \tag{6.27}$$

(partitioning of sulphur at the slag–metal interface).

It is found that approximately 15–25% of dissolved sulphur is directly oxidized (Equation (6.26)) into the gaseous phase due to the turbulent and oxidizing conditions existing in the jet impact zone.

The rate equation for sulphur transfer assuming metal phase control was derived in Example 3.2. In a steel plant, regression equations based on operational data are employed to predict the end point sulphur within acceptable limits. One such equation, for example [26], is

$$\frac{(\%S)}{[S]} = 1.42\,B - 0.13(\%\,FeO) + 0.89 \tag{6.28}$$

This equation shows the beneficial influence of slag basicity B and the retarding influence of FeO on sulphur distribution. A large number of such correlations are reported in the literature, but they are suitable and applicable to a local situation only. Regression equations based on optical basicity have also been proposed to predict end point sulphur (see Chapter 2).

6.7   Kinetic model for slag–metal–gas reactions in the presence of an oxygen jet

The reactions occurring during steelmaking form a good example of a multi-component–multiphase system involving slag, metal and gas. Multicomponent mixed transport theory has been developed [27] to describe simultaneous oxidation/reduction reactions in a slag/metal/gas system. A fairly involved dynamic model, which incorporates the effects of changes in bath temperature on the reaction kinetics, has also been developed and applied to the 300 ton converter at Hoogovens IJmuiden [28, 29]; typical results of the model are shown in Figure 6.12. In the example given below, only the oxidation rates of carbon, silicon, iron and manganese are considered for the sake of simplicity.

Example 6.1 **Dynamic model of the simultaneous oxidation of carbon, silicon, manganese and iron in a slag–metal–oxygen jet system adapted from Deo *et al.* [28]**

It can be assumed that an instantaneous thermodynamic equilibrium exists at various metal–slag–gas interfaces and that the rates of reactions are governed by the transport of species, to or away from the interface. The calculation can be done in several steps.

Step 1: Calculation of thermodynamic equilibrium at the interface

The concentration of species present is obtained from equilibria at the interface, i.e. for the reaction

$$\{Fe\} + [O] = (FeO) \tag{A}$$

The equilibrium constant is

$$K_1 = \frac{\gamma_{FeO} X^i_{FeO}}{f_O X^i_O X^i_{Fe}} \tag{B}$$

X^i signifies the interfacial mole fraction of different species of FeO, iron and oxygen at the slag–metal interface, γ_{FeO} is the activity coefficient of FeO and f_O is the Henrian activity coefficient for oxygen at infinite dilution. The mole fraction of FeO

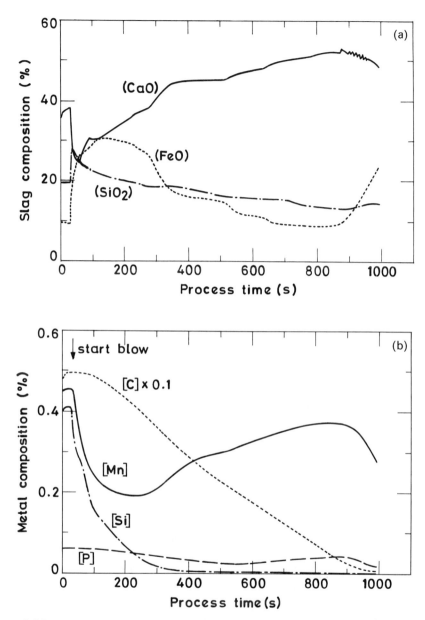

Figure 6.12 Predicted composition of slag and metal by dynamic slag–droplet model:
(a) slag composition; (b) metal composition.

at the interface can be calculated from

$$X^i_{FeO} = kOX_1 X^i_O X^i_{Fe} \tag{C}$$

where $kOX_1 = K_1 f_O/\gamma_{FeO}$.

For silicon, the equations are

$$[Si] + 2[O] = (SiO_2) \tag{D}$$

$$K_2 = \frac{\gamma_{SiO_2} X^i_{SiO_2}}{X^i_{Si} f_{Si} f^2_O (X^i_O)^2} \tag{E}$$

and

$$X^i_{SiO_2} = kOX_2 (X^i_O)^2 X^i_{Si} \tag{F}$$

where

$$kOX_2 = (K_2 f^2_O f_{Si})(1/\gamma_{SiO_2}) \tag{G}$$

For manganese

$$[Mn] + [O] = (MnO) \tag{H}$$

$$K_3 = \frac{\gamma_{MnO} X^i_{MnO}}{f_O X^i_O f_{Mn} X^i_{Mn}} \tag{I}$$

and

$$X^i_{MnO} = kOX_3 X^i_O X^i_{Mn} \tag{J}$$

where

$$kOX_3 = K_3 f_O f_{Mn}/\gamma_{MnO} \tag{K}$$

Step 2: Flux density equations

The transfer of elements takes place mainly across slag–metal, slag–gas and gas–metal interfaces. Let X^b denote the bulk mole fraction of each element. The flux density equation for each element is obtained from mass balance. For the transfer of iron across the slag–metal interface, the rate of transfer out of the metal equals the rate of transfer into the slag, or

$$k_m C_{Vm}(X^b_{Fe} - X^i_{Fe}) = k_s C_{Vs}(X^i_{FeO} - X^b_{FeO}) \tag{L}$$

where k_m and k_s are the mass transfer coefficients in metal and slag, respectively, C_{Vm} and C_{Vs} are the molar densities of metal and slag, respectively.

Let $k_s C_{Vs}/k_m C_{Vm} = k^*$ and $TM_1 = X^b_{Fe} + k^* X^b_{FeO}$. Then

$$X^i_{Fe} = TM_1/(1 + k^* k\theta X_1 X^i_O) \tag{M}$$

For the transfer of silicon

$$k_m C_{Vm}(X_{Si}^b - X_{Si}^i) = k_s C_{Vs}(X_{SiO_2}^i - X_{SiO_2}^b)$$

If $TM_2 = X_{Si}^b + k*X_{SiO_2}^b$, then

$$X_{Si}^i = TM_2/[1+k*kOX_2(X_O^i)^2]$$

Similarly, for the transfer of manganese, if

$$TM_3 = X_{Mn}^b + k*X_{MnO}^b,$$

then

$$X_{Mn}^i = TM_3/(1+k*kOX_3 X_O^i)$$

Thus, interfacial concentrations of silicon, manganese and iron are obtained in terms of their bulk concentrations, transport properties and interfacial oxygen mole fraction. For oxygen,

$$k_m C_{Vm}(X_O^b - X_O^i) = k_s C_{Vs}[(X_{FeO}^i - X_{FeO}^b) + 2(X_{SiO_2}^i - X_{SiO_2}^b)$$

$$+ (X_{MnO}^i - X_{MnO}^b)] + \text{carbon oxidation term} \qquad \text{(N)}$$

Or

$$X_O^b - X_O^i = k*(A_1 + A_2 + A_3) + \text{carbon oxidation term} \qquad \text{(O)}$$

where:

$$A_1 = X_{FeO}^i - X_{FeO}^b = kOX_1 X_O^i TM_1/(1 + k*kOX_1 X_O^i) - X_{FeO}^b$$

$$A_2 = 2[kOX_2 TM_2(X_O^i)^2]/[1 + k*kOX_2(X_O^i)^2] - 2X_{SiO_2}^b$$

$$A_3 = X_{MnO}^i - X_{MnO}^b = kOX_3 X_O^i TM_3/(1 + k*kOX_3 X_O^i) - X_{MnO}^b$$

The non-linear equation (O) in one unknown X_O^i can be solved by the Newton–Raphson method, provided that the carbon oxidation term is known.

Step 3: Equations for the oxidation of carbon

The bubble of carbon monoxide at the slag–gas interface reduces some metal oxides in the slag phase:

$$(MO) + (CO)_g = [M] + (CO_2)_g \qquad \text{(P)}$$

The carbon dioxide gas thus formed is transported to the gas–metal interface where it is reduced by metal and carbon components:

$$(CO_2)_g + [M] = (CO)_g + (MO)$$

$$(CO_2)_g + [C] = 2(CO)_g$$

Again:

$$(MO) = [M] + [O] \qquad \text{(Q)}$$

Thus, the effect of a bubble containing carbon monoxide and carbon dioxide gases is to increase the oxygen flux from the slag to the metal. This effect of bubbles can be incorporated by way of the oxygen flux parameter F_O. Consider the reaction

$$[C] + [O] = (CO)_g \tag{R}$$

for which

$$K_4 = \frac{p_{CO}}{X_C^i X_O^i f_O \gamma_C} \tag{S}$$

where γ_C is the Raoultian activity coefficient of carbon. Let

$$kOX_4 = K_4 f_O \gamma_C \tag{T}$$

From mass balance, assuming for simplicity that carbon is first oxidized to carbon monoxide, the rate of carbon consumption equals the rate of carbon monoxide evolution, or

$$-\frac{dC}{dt} = \frac{dp_{CO}}{dt} \tag{U}$$

But $dC/dt = k_m C_{vm}(X_C^b - X_C^i)$ and if G_{CO} is the parameter for carbon monoxide evolution, then

$$k_m C_{vm}(X_C^b - X_C^i) = -G_{CO}(p_{CO} - 1) \tag{V}$$

From Equations (S) and (T)

$$p_{CO} = kOX_4 X_O^i X_C^i \tag{W}$$

If we let:

$$\frac{G_{CO}}{k_m C_m} = G \quad \text{(dimensionless)} \tag{X}$$

then

$$X_C^b - X_C^i = -G(kOX_4 X_O^i X_C^i - 1) \tag{Y}$$

and

$$X_C^i = TM_4/(1 - GkOX_4 X_O^i) \tag{Z}$$

where

$$TM_4 = X_C^b - G \tag{A1}$$

Introducing a parameter F_O such that

$$F_O k_m C_{vm}(X_O^b - X_O^i) = k_s C_{vs}[(X_{FeO}^i - X_{FeO}^b) + (X_{MnO}^i - X_{MnO}^b)$$
$$+ 2(X_{SiO_2}^i - X_{SiO_2}^b)] + G_{CO}(p_{CO} - 1) \tag{A2}$$

i.e.

$$(X_O^b - X_O^i) = \frac{k*}{F_O}[(X_{FeO}^i - X_{FeO}^b) + (X_{MnO}^i - X_{MnO}^b) + 2(X_{SiO_2}^i - X_{SiO_2}^b)]$$

$$+ \frac{G_{CO}}{k_m C_{Vm} F_O}(p_{CO} - 1) \tag{A3}$$

$$= \frac{k*}{F_O}(A_1 + A_2 + A_3) + \frac{G}{F_O}A_4 \tag{A4}$$

where

$$A_4 = kOX_4 X_O^i TM_4/(1 - GkOX_4 X_O^i) - 1 \tag{A5}$$

Equation (A5) can be solved for X_O^i by the Newton–Raphson method.

Step 4: Conservation equations to modify bulk metal, gas and slag compositions

Knowing all the interfacial concentrations, we can calculate the rates at which bulk concentrations will change. The conservation equation for iron is

$$C_{Vm} V_m \frac{\Delta X_{Fe}^b}{\Delta t} = -Ak_m C_{Vm}(X_{Fe}^b - X_{Fe}^i) \tag{A6}$$

where V_m is the volume of metal and A is the interfacial area.

At each time step Δt a mass balance is carried out and the molar compositions of slag and metal phases are modified, depending upon the amount of transfer that might have taken place. The heat (or energy) balance is incorporated at each time step to consider the heat effects of reactions and also heat loss from the system so that in the subsequent time step the modified temperatures of slag, gas and metal phases are used for the calculation of equilibria at the interfaces. A modified oxygen balance is used to account for oxygen supplied externally by the lance ($\Delta M/\Delta t$) as follows:

$$F_O k_m (X_O^b - X_O^i)A + \frac{\Delta M}{\Delta t}$$

$$= k_s C_{Vs} A[(X_{FeO}^i - X_{FeO}^b) + (X_{MnO}^i - X_{MnO}^b) + 2(X_{SiO_2}^i - X_{SiO_2}^b)]$$

$$+ \text{carbon oxidation term} \tag{A7}$$

The actual values of $\Delta M/\Delta t$ are calculated at each time step. The calculations can be made more representative of the actual system by incorporating the rates of scrap dissolution, lime dissolution, ore dissolution, emulsification and droplet generation.

The solubility and activity of different components in steelmaking slags can be calculated from the Kapoor–Frohberg model (described in Chapter 2). The mass transfer coefficient for the dissolution of lime in slag is found by conducting separate experiments. A similar exercise is done for ore dissolution.

Metal droplets are ejected from the jet impact region. The size distribution of these droplets (see Chapter 5) depends on nozzle design and process parameters. The amount of droplets generated can also be related to process parameters (see Chapter 5 on slag formation). The CO–CO_2 gases, when entrapped in slag together with the droplets, lead to the formation of a gas–metal–slag emulsion. The reactions occurring in the emulsion are complex and may depend on lime dissolution, slag composition, viscosity, temperature, gas evolution rate, intensity of blowing, lance type, height, etc. Therefore, emulsion formation behaviour is specific to a process and empirical correlations are needed to model the system accurately. Details of such calculations are provided in a recent work [29].

Step 5: Heat balance and the evaluation of new bath temperature at each time step

The aim of heat balance (or energy balance) calculations is to determine the temperature of the bath from the heat evolved by reactions within the system and the heat exchange between the system and its surroundings. These calculations depend upon the size and shape of the vessel used. As a simple example of heat balance, the heat equivalent of silicon in hot metal is calculated in Example 6.2 below.

Example 6.2 Heat equivalent of silicon in hot metal

It is required to estimate the temperature rise associated with the oxidation of 0.6 wt % silicon from blast furnace iron. Given that the initial temperature of iron is 1673 K and the temperature of oxygen is 298 K:

$$\langle Si \rangle + (O_2)_g = \langle SiO_2 \rangle; \qquad \Delta H_{298} = -905\,000\,\text{J/mol}$$

$$[Si](1 \text{ wt \% in liquid Fe}) = \{Si\}; \qquad \Delta H_{1673} = +119\,000\,\text{J/mol}$$

$$2\langle CaO \rangle + \langle SiO_2 \rangle = \langle Ca_2SiO_4 \rangle; \qquad \Delta H_{298} = -126\,000\,\text{J/mol}$$

$$\{Si\} = \langle Si \rangle; \qquad \Delta H_{1673} = -46\,000\,\text{J/mol}$$

$$\left.\begin{array}{l}\langle Ca_2SiO_4 \rangle_{298\,K} = \langle Ca_2SiO_4 \rangle_{1673\,K}; \quad \Delta H_5 \\[4pt] 2\langle CaO \rangle_{1673\,K} = 2\langle CaO \rangle_{298\,K}; \quad\quad\ \Delta H_6 \\[4pt] \langle Si \rangle_{1673\,K} = \langle Si \rangle_{298\,K}; \quad\quad\quad\quad\ \Delta H_7\end{array}\right\} \text{information needed}$$

$$O_2(g, 298\,K) + Si(1 \text{ wt \% in liquid Fe}, 1673\,K) + 2\langle CaO \rangle_{1673K}$$

$$= \langle Ca_2SiO_4 \rangle_{1673K}; \Delta H_8$$

$$C_p \langle Ca_2SiO_4 \rangle = 151 + 36.8 \times 10^{-3}T - 30.4 \times 10^5 T^{-2} \text{ J/K mol}$$

$$C_p\, 2\langle CaO \rangle \quad = 99.0 + 9.00 \times 10^{-3}T - 13.9 \times 10^5 T^{-2} \text{ J/K mol}$$

$$C_p \langle Si \rangle \quad\quad = 24.0 - 2.46 \times 10^{-3}T - 4.13 \times 10^5 T^{-2} \text{ J/K mol}$$

Assumptions

The heat associated with the incorporation of silica into the slag may be approximated by the heat of formation of dicalcium silicate:

$$\Delta H_5 + \Delta H_6 + \Delta H_7 = \int_{298}^{1673} (151 + 36.8 \times 10^{-3}T - 30.4 \times 10^5 T^{-2})\, dT$$

$$- \int_{298}^{1673} (99.0 + 9.00 \times 10^{-3}T - 13.9 \times 10^5 T^{-2})\, dT$$

$$- \int_{298}^{1673} (24.0 - 2.46 \times 10^{-3}T - 4.13 \times 10^5 T^{-2})\, dT$$

$$= \int_{298}^{1673} \Delta C_p\, dT$$

$$151 + 36.8 \times 10^{-3}T - 30.4 \times 10^5 T^{-2}$$

$$-99.0 - 9.00 \times 10^{-3}T + 13.9 \times 10^5 T^{-2}$$

$$\underline{-24.0 + 2.46 \times 10^{-3}T + 4.13 \times 10^5 T^{-2}}$$

$$\Delta C_p = 28.0 + 30.26 \times 10^{-3}T - 12.37 \times 10^5 T^{-2}$$

$$\Delta H_5 + \Delta H_6 + \Delta H_7 = 28.0(1673 - 298) + \frac{30.26}{2} \times 10^{-3}(1673^2 - 298^2)$$

$$+ 12.37 \times 10^5 \left(\frac{1}{1673} - \frac{1}{298} \right)$$

$$= 38\,500 + 41\,004 - 3412$$

$$= 76\,092\ \text{J/mol}$$

$$\Delta H_8 = -905\,000 + 119\,000 - 126\,000 - 46\,000 + 76\,092$$

$$= -881\,908\ \text{J/g mol of silicon}$$

i.e. exothermic and leading to a temperature rise. Now consider 1 g of blast furnace iron; this contains $(0.6 \times 10^{-2})/28.06$ moles of silicon.

If the specific heat of the iron is $0.832\ \text{J/g K}$ the temperature rise (ΔT) under adiabatic conditions is given by

$$\Delta T = \frac{0.6 \times 10^{-2} \times 881\,908}{28.06 \times 0.832}$$

$$= 226.7\ \text{K}$$

Similar calculations can be done for the oxidation of phosphorus, manganese and carbon dissolved in iron. If heat losses to the ambient atmosphere by radiation, convection and conduction are known then an input–output heat balance can be made for the converter. Static charge balance models are thus developed to predict rough estimates of the hot metal, scrap, ore, lime and oxygen gas (and hence blow time) required for a particular heat in advance. The dynamic model can then be used to predict the course of reactions and the final temperature and composition of steel at the end of the blow. The literature is replete with examples of static and dynamic control models of converters. In fact control models for specific applications can be easily purchased on a commercial basis.

References

1. Lange, K. W. (1981) *Physical and Chemical Influences during Mass Transfer in Oxygen Steelmaking Process*, West Deutscher Verlag: Lingerich.
2. Schürmann, E., Sperl, H., Hammer, R. and Ender, A. (1986) 'Metallurgische Interpretation der CO_2-Gehalte des Konvertergases beim Sauerstoff aufblasverfahren' *Stahl und Eisen*, **106**, 1267–72.
3. Koch, K., Fix, W. and Valentin, P. (1978) 'Einfluss von Sauerstoffangebot und Kohlenstoffausgangsgehalt sowie von Badgeometrie und Feuerfestmaterial auf den Ablauf der Entwicklung von Fe-C-Schmelzen in einem 50-kg-Aufblastkonverter' *Arch. Eisenhüttenwes.*, **49**, 231–4.
4. Koch, K., Fix, W. and Valentin, P. (1976) 'Einsatz eines 50-kg-Aufblastkonverters zur Untersuchung der Entkohlung von Eisen-Kohlenstoff-Schmelzen' *Arch. Eisenhüttenwes.*, **47**, 659–63.
5. Schürmann, E., Florin, W. and Sperl, H. (1987) 'Phosphorus and carbon oxidation in converter', *Stahl und Eisen*, **107**, 1091–7.
6. Koch, K., Fix, W. and Valentin, P. (1978) 'Entkohlungsreaktionen mit unruhigem Blasverhalten beim Aufblasten von Sauerstoff auf Fe-C-Schmelzen' *Arch. Eisenhüttenwes.*, **49**, 163–6.
7. Koch, K., Fix, W. and Valentin, P. (1978) 'Kennzeichende Teilabschnitte der Entkohlungs-reaktion beim O_2-Aufblasen auf Fe-C-Schmelzen' *Arch. Eisenhüttenwes.*, **49**, 109–14.
8. Van Unen, G., van der Knoop, W., Mink, P., Snoeijer, A. B. and Boom, R. (1991) 'Use of the sublance to study the BOS process', *NSC-Hoogovens Sublance Conf.*, September *1991*, pp. 1–10.
9. Gaye, H. and Riboud, P. V. (1977) 'Oxidation kinetics of iron alloy drops in oxidizing slags', *Metall. Trans.*, **8B**, 409–15.
10. Rao, Y. K. and Lee, H. G. (1988) 'Decarburisation and nitrogen absorption in molten Fe–C alloys: Part 1 Experimental', *Ironmaking and Steelmaking*, **15**, 228–37.
11. Lee, H. G. and Rao, Y. K. (1988) 'Decarburisation and nitrogen absorption in molten Fe–C alloys: Part 2 Mathematical model', *Ironmaking and Steelmaking*, **15**, 238–43.
12. Sain, D. R. and Belton, G. R. (1976) 'Interfacial reaction kinetics in the decarburization of liquid iron by carbon dioxide', *Metall. Trans.*, **7B**, 235–44.
13. Lee, H. G. and Rao, Y. K. (1982) 'Rate of decarburization of iron–carbon melts: Part II. A mixed-control model', *Metall. Trans.*, **13B**, 411–21.
14. Greenberg, L. A. and McLean, A. (1974) 'The kinetics of oxygen dissolution in liquid iron and liquid iron alloy droplets', *Trans. Iron Steel Inst. Jpn*, **14**, 395–403.

15. Baker, L. A., Warner, N. A. and Jenkins, A. E. (1967) 'Decarburization of a levitated iron carbon droplet in oxygen', *Trans. Met. Soc. AIME*, **239**, 857–64.

16. Lee, H. G. and Rao, Y. K. (1982) 'Rate of decarburization of iron–carbon melts: Part 1. Experimental determination of the effect of sulfur', *Metall. Trans.*, **13B**, 403–9.

17. Ito, K., Amano, K. and Sakao, H. (1984) 'Kinetics of carbon- and oxygen-transfer between CO–CO_2 mixture and molten iron', *Trans. Iron Steel Inst. Jpn*, **24**, 515–21.

18. Fruehan, R. J. and Antolin, S. (1987) 'A study of the reaction of CO on liquid iron alloys', *Metall. Trans.*, **18B**, 415–20.

19. Sommerville, I. D., Grieveson, P. and Taylor, J. (1980) 'Kinetics of reduction of iron oxide in slag by carbon in iron: Part 1 effect of oxide concentration', *Ironmaking and Steelmaking*, **7**, 25–32.

20. Fine, H. A., Meyer, D., Janke, D. and Engell, H. J. (1985) 'Kinetics of reduction of iron oxide in molten slag by CO at 1873 K', *Ironmaking and Steelmaking*, **12**, 157–62.

21. Mroz, J. (1987) 'Reduction of iron oxides from liquid ferrous slags by blast furnace coke', *Scand. J. Met.*, **16**, 16–22.

22. Sato, A., Aragane, G., Hirose, F., Nakagawa, R. and Yoshimatsu, S. (1984) 'Reducing rate of iron oxide in molten slag by carbon in molten iron', *Trans. Iron Steel Inst. Jpn*, **24**, 808–15.

23. Turkdogan, E. T. (1984) 'Physicochemical aspects of reactions in ironmaking and steelmaking processes', *Trans. Iron Steel Inst. Jpn*, **24**, 591–611.

24. Bergman, A. (1990) 'The application of optical basicity to dephosphorization equilibria', *Steel Res.*, **61**, 347–52.

25. Ono, H., Inagaki, A., Masui, T., Narita, H., Nosaka, S., Mitsuo, T. and Gohda, S. (1981) 'Removal of phosphorus from LD converter slag by floating separation of dicalcium silicate during solidification', *Trans. Iron Steel Inst. Jpn*, **21**, 135–44.

26. Deo, B. (1988) 'Models for predicting sulphur and phosphorus distribution ratios in oxygen steelmaking', *Trans. Indian Inst. Met.*, **41**, 475–9.

27. Robertson, D. G. C., Deo, B. and Ohguchi, S. (1984) 'Multicomponent mixed-transport-control theory for the kinetics of coupled slag/metal and slag/metal/gas reactions: application to desulphurization of molten iron', *Ironmaking and Steelmaking*, **11**, 41–55.

28. Deo, B., Ranjan, P. and Kumar, A. (1987) 'Mathematical model for computer simulation and control of steelmaking', *Steel Res.*, **58**, 427–31.

29. Van der Knoop, W., Deo, B., Snoeijer, A. B., van Unen, G. and Boom, R. (1992) 'A dynamic slag–droplet model for the steelmaking process', *4th Int. Conf. on Molten Slags and Fluxes, 1992, Sendai, Japan, Proc. Trans. Iron Steel Inst. Jpn, pp. 302–7.*

7 Secondary steelmaking metallurgy

7.1 Treatment of liquid steel in ladle

The process of transferring liquid steel from the furnace to the ladle is called tapping. Towards the end of tapping some amount of furnace slag is carried into the ladle along with the metal. This slag is known as 'carried over slag'. Depending upon the particular process route followed at a plant, deoxidation and alloying additions may be made during tapping. Intense agitation of the bath caused by the falling stream of liquid metal in the ladle helps in alloy dissolution, deoxidation and mixing (i.e. bath homogenization). Loss of deoxidizing elements added in the ladle (namely, loss of aluminium and silicon) due to reaction with FeO, MnO, SiO_2 and P_2O_5 in the carried over slag depends upon the amount and chemical composition of the carried over slag [1]. Since the alloying elements are expensive, every effort is made to minimize their loss (or, in other words, maximize their recovery). Further, when the reaction products, like Al_2O_3 and SiO_2, are entrapped in the metal, then the cleanliness of steel in the ladle may be adversely affected. The reduction of P_2O_5 in carried over slag by dissolved aluminium and silicon present in the metal can lead to a rise in the phosphorus content of the metal (or phosphorus reversion); strict control of phosphorus is desirable so as to avoid grain boundary embrittlement in the product. Carried over slag also influences, though to a limited extent, the hydrogen and nitrogen pickup by liquid steel during or after tapping. Owing to these reasons, an attempt is made to minimize slag carryover from the furnace to the ladle.

As a first step to minimize slag carryover, slag stoppers are placed on the tap holes of the furnace towards the end of the tapping operation. Once the steel has been tapped in the ladle the carried over slag can be removed by mechanical raking, or by ladle-to-ladle transfer or by employing a vacuum slag cleaner. All these operations are, however, expensive.

In order to improve the quality of liquid steel in the ladle various treatments are possible:

- *Stirring treatment with argon gas* can be done to homogenize the bath and to promote decarburization (by the reaction of dissolved carbon and oxygen due to the low partial pressure of carbon monoxide gas in the argon bubbles) and degassing (due to the low partial pressure of hydrogen and nitrogen in the argon bubbles). Alloy dissolution, deoxidation and slag–metal reactions are also enhanced due to the stirring effect.

- *Exchange treatment with synthetic slag* of desired composition (low in FeO) can be done by adding a suitable slag to the ladle before or during tapping in order to achieve deoxidation, desulphurization and inclusion modification [2]. During the treatment with synthetic slag, argon stirring is usually done to enhance the slag–metal reaction.
- *Exchange treatment with conditioned slag* can be done (namely, by adding pure CaO or a mixture of limestone and aluminium) to the top slag in order to lower the activity of FeO in the slag [3]. Even the synthetic slag can be conditioned.
- *Vacuum treatment* can be done to facilitate degassing and enhance the kinetics of decarburization and deoxidation reactions. Alloying additions can be made to adjust the composition during the treatment. Argon stirring is an integral part of most of the vacuum treatments.
- *Injection treatment* can be done by injecting inert gas, inert gas plus oxygen, or pure oxygen, or by injecting powdered agents like CaSi, CaC_2, CaAl, rare earth metal and magnesium into the metal with the help of an inert carrier gas, or by submerged cored-wire injection. In some cases a prior treatment with synthetic slag is a prerequisite for injection treatment. This is because in the absence of synthetic slag both the recovery and the efficiency of injection treatment decreases. During injection treatment deoxidation, desulphurization and inclusion modification can occur simultaneously. These reactions are usually classified as precipitation-type reactions. Subsequent purging with argon can help in bath homogenization and oxide flotation and hence in the production of clean steel. Powders of materials like CaO, $CaO–CaF_2$ and $CaO–Al_2O_3–CaF_2$ can be injected for desulphurization. Alloy powders can also be injected to adjust the composition.

The variety of treatments described above can be suitably placed into two main process routes for the production of clean steel, namely, the argon-stirring route and the vacuum treatment route:

1. In the argon-stirring route, synthetic slag may be added at the time of furnace tapping to achieve desulphurization. Slag conditioning can also be done at the time of tapping. After deoxidation at tapping, steel is homogenized at the stirring station by passing argon through a porous plug or through a submerged injection lance. Alloying, desulphurization and deoxidation additions to the steel can be made by wire injection followed by soft argon purging to help in the flotation of reaction products.
2. In the vacuum treatment route, if the objective is to produce ultra-low carbon grades, namely by treatment in RH (or RH-OB), then deoxidation is not done at the time of tapping because dissolved oxygen is required later to react with

Table 7.1 Average slag compositions in wt % at the start (ST) and end (ET) of ladle treatment for ultra-low carbon grades (RH-OB route) and tinplate grades (argon-stirring route).

wt %	RH-OB route		Argon-stirring route	
	ST	ET	ST	ET
FeO	16	<20	5	4
MnO	6	5–10	3	3
Al_2O_3	18	32	40	40
SiO_2	11	10	7	9
CaO	43	33	41	40

carbon. Alloying additions, depending upon their nature, can be made at a suitable stage during the course of vacuum treatment; for instance, the ferro alloys (like Fe–Nb) which are susceptible to loss by oxidation are added after deoxidation.

Injection treatment can be done at a convenient stage in the argon-stirring route or the vacuum treatment route. In fact, although injection metallurgy is in principle an integral part of secondary steelmaking metallurgy, it is now regarded as a separate topic within the broad area of steelmaking. Hence the fundamentals of injection metallurgy will be discussed in Chapter 8.

Typical slag compositions for the vacuum treatment route (i.e. incorporating RH-OB but without slag conditioning) and argon-stirring route at the start (ST) and end (ET) of ladle treatment (without slag conditioning) are given in Table 7.1. The ladle slag has a higher Al_2O_3 content in the argon-stirring route because in this case deoxidation is carried out during tapping; aluminium reacts with oxygen to form Al_2O_3.

The fundamental considerations for slag–metal reactions based on thermo-dynamics and the kinetics of reactions in a ladle during (a) furnace tapping, (b) argon stirring and vacuum treatment, (c) exchange treatment with synthetic slag, and (d) exchange treatment with conditioned slag are described in sections 7.2 to 7.5. Deoxidation and decarburization under reduced pressure of carbon monoxide gas are described in section 7.6. Degassing during argon stirring and vacuum treatment are described in section 7.7.

7.2 Reaction between carried over slag and metal during furnace tapping

The converter slag at the time of tapping is highly oxidizing in nature because it may contain 12–20% FeO. The reaction of FeO in the furnace slag with dissolved aluminium, silicon and manganese in metal is primarily responsible for the loss (or low recovery) of deoxidants added in the ladle; for low-carbon heats the percentage recovery of silicon is approximately 70% (i.e. 30% of the mass of silicon is lost into slag as SiO_2); the recovery of aluminium may range from 35 to 65%. Phosphorus reversion depends upon the activity of P_2O_5 in the slag and aluminium and silicon content of the metal. If the amount of carried over slag is approximately known (from the visual estimation of slag thickness) then it is possible to do simple material balance calculations and develop regression relations on the basis of plant data. It is not worthwhile (as also demonstrated later in the analysis of plant data in slag conditioning) to develop accurate thermodynamic and kinetic models owing to the uncertainties involved. Fundamental aspects of the reactions between carried over slag and liquid steel during furnace tapping, as discussed below, have been adapted from Turkdogan [1].

7.2.1 *Material balance for aluminium, silicon and phosphorus*

Let the amount of aluminium added to the bath be Al_{add}; then from material balance

$Al_{add} = $ Al oxidized by furnace slag

+ Al oxidized due to air entrainment

+ Al dissolved in steel

+ Al reacted with dissolved oxygen and retained as Al_2O_3 inclusions in steel

+ Al reacted with dissolved oxygen and floated up as Al_2O_3

The loss of aluminium due to reaction with oxygen picked up by the metal (from the entrainment of air) is negligible compared with that lost by reaction with FeO in the furnace slag. Oxygen pick-up due to air entrainment is usually less than 20 ppm. The oxygen dissolved in steel can be determined by sublance measurement before tapping the steel in the ladle. Thus, by systematically analyzing the slag and metal and doing material balance for an estimated amount of slag carried over to the ladle, each shop can develop regression relations to predict the overall recovery of aluminium.

The material balance for silicon (in the case of aluminium-killed steels) can be written as

Si added to metal = Si oxidized by FeO, MnO and P_2O_5 in furnace slag

+ Si dissolved in steel

For a 200 ton tap ladle the following approximate relation is proposed [1] for the combined loss of aluminium and silicon:

% Al + % Si oxidized by furnace slag = 1.1×10^{-6} (% change in FeO and MnO of slag) × slag weight + 1.1×10^{-4} (weight of skull in ladle plus fallen converter skull from furnace)   (7.1)

Skull is solidified steel plus slag mass sticking to the ladle walls plus skull fallen from the mouth of the converter.

Suppose the percentage change in the FeO and MnO content of slag is 5%, the weight of carried over slag is 3000 kg, and the weight of skull in the ladle is 2000 kg. Then % Al + % Si oxidized by furnace slag in a 200 ton ladle will be 0.0715%.

The empirical equation for phosphorus reversion to metal is [1]

$$\Delta P \text{ (wt \%)} = 0.05 \times (\% \text{ change in } P_2O_5 \text{ content of slag}) \times \text{slag weight} \qquad (7.2)$$

7.3 Reaction between carried over slag and metal during argon stirring and vacuum treatment

In this section the kinetics of the oxidation of dissolved aluminium by top slag in an argon-stirred ladle and during RH vacuum treatment is considered. The Al_2O_3 thus formed greatly affects the cleanliness of the steel in the ladle.

7.3.1 Kinetics of oxidation of dissolved aluminium by carried over slag in an argon-stirred ladle

The effect of stirring induced by inert gas injection on mixing times (bath homogenization) and on the mass transfer coefficient in the metal phase has been discussed in Chapter 1. The thermodynamics of slag–metal reactions and the kinetics of reactions in a gas-stirred bath, including deoxidation and desulphurization, have been discussed in Chapters 2 and 3.

The slag carried over from the oxygen steelmaking furnace to the ladle may contain 12 to 17 wt % FeO (see Table 7.1). The change in soluble aluminium (Al_{sol}) in metal and in the iron content of slag is plotted as a function of time in Figure 7.1 for a 185 ton high-alumina ladle covered with 1640 kg or 410 kg of slag and stirred by 0.75 m³ (stp)/min of argon gas [4]; other conditions remaining the same, the loss of Al_{sol} increases if the amount of slag increases (Figure 7.1).

If it is assumed that the mass transfer of aluminium in the metal is rate limiting then, similar to Equation (A) in Example 3.2, we can write:

$$-\ln\left(\frac{Al_t - Al_e}{Al_o - Al_e}\right) = k_{Al}\frac{A}{V}t$$

where Al_t is wt % aluminium in metal at any time t, Al_e is wt % aluminium at equilibrium, Al_o is wt % aluminium at start, k_{Al} is the mass transfer coefficient of aluminium in metal (m/s), A is the nominal interface area (m²) and V is the volume of metal (m³). Since Al_e is much smaller than Al_t and Al_o, the above equation reduces to

$$-\ln\left(\frac{Al_t}{Al_o}\right) = k_{Al}\frac{A}{V}t$$

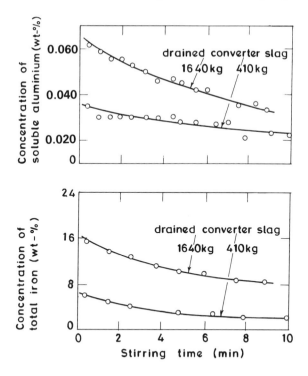

Figure 7.1 Oxidation of soluble aluminium in metal due to reaction with 1640 kg and 410 kg top slag in a 185 ton ladle stirred with gas. (*Source*: Pluschkell, W., Redenz, B. and Schürmann, E. (1981), *Arch. Eisenhüttenwes.*, **52**, 82–90.)

Thus, by plotting $\ln(Al_t/Al_o)$ versus time we can determine the mass transfer coefficient k_{Al} for given values of A and V. As a rough estimate from Figure 7.1, if $Al_o = 0.06$ wt %, $Al_t = 0.04$ wt % at 360 s, $A = 7\,\text{m}^2$ for the top slag–metal interface and $V = 26.4\,\text{m}^3$ for a 185 ton ladle, then $k_{Al} = 4.2 \times 10^{-4}\,\text{m/s}$.

The assumption made a priori that the mass transfer of aluminium is rate controlling only in the metal phase is an oversimplification and is valid only for slags rich in FeO. It will be shown later that for slags impoverished in FeO (see section 7.4 which is devoted to treatment with non-oxidizing synthetic slags) the mass transfer of oxygen in slag becomes rate controlling. A kinetic model assuming mixed transport control in metal as well as in slag is developed in Example 7.1.

The plot of $\lg(Al_t/Al_o)$ versus time for a 100 ton ladle is shown in Figure 7.2. For the typical conditions [5] of 100 tons of metal, 30.3 m² total surface area (wall plus slag–metal interface), 14.3 m³ metal volume stirred by approximately 0.4 m³ (stp)/min argon gas, the value of k_{Al} is approximately $6.4 \times 10^{-4}\,\text{m/s}$. The scatter in Figure 7.2 is larger than in Figure 7.1 due to a greater degree of turbulence and instability of the slag–metal interface in a smaller ladle.

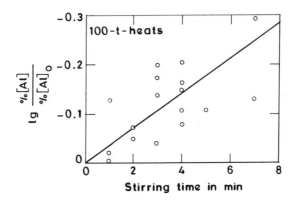

Figure 7.2   Rate law for the decrease of aluminium content of steel during gas stirring in ladle with 100 ton heats. (*Source*: Karl, K. H., Abratis, H., Maas, H. and Wahlster, M. (1974), *Arch. Eisenhüttenwes.*, **45**, 9–16.)

A theoretical model of the effect of slag droplet entrapment and the existence of dead (completely unmixed) zones on the mass transfer process in a gas-stirred bath has been presented by Mietz and Bruhl [6]; the dead zones and the emulsified droplets are schematically shown in Figure 7.3. Such a model has several adjustment factors which can, in principle, be tuned to a given set of conditions to describe the process well and hence determine the optimum design of the vessel and the operating conditions. According to the results of cold model experiments [6], the volume of dead zone in a ladle can increase from 17 to 35% if the gas flow is decreased. The contribution of the emulsification of slag and metal near the slag–metal interface to the overall reaction kinetics can introduce many fluctuations in the predicted results from one situation to the other.

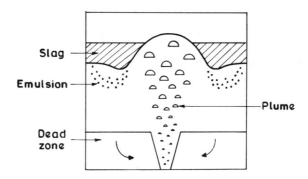

Figure 7.3   Emulsification of slag droplets and existence of dead zones in an argon-stirred ladle. (*Source*: Mietz, J. and Brühl, M. (1990), *Steel Res.*, **61**, 105–12.)

Table 7.2 Mass transfer coefficients for aluminium, k_{Al}, during RH process and argon-stirring treatment.

Treatment	Capacity (tons)	Mass transfer coefficient (m/s)	Ref.
RH	35	3.7×10^{-4}	[5]
	100	2.5×10^{-4}	[5]
	200	1.3×10^{-4}	[5]
Argon stirring	100	6.4×10^{-4}	[5]
	185	4.2×10^{-4}	[4]

7.3.2 *Kinetics of oxidation of dissolved aluminium by carried over slag in RH vessels*

Detailed analysis of the kinetics of aluminium oxidation in RH vessels of various sizes has been reported in the literature [5]. The calculated mass transfer coefficients are summarized in Table 7.2; the orders of magnitude of the mass transfer coefficients obtained for vessels of different sizes are similar to those for argon-stirring treatment. Smaller vessels (compare data for the 35 ton and 200 ton units in Table 7.2) exhibit a higher mass transfer coefficient for aluminium in liquid metal and therefore a higher rate of oxidation of aluminium by the top slag.

7.4 Exchange treatment with non-oxidizing synthetic slags

The concept of using synthetic slags in the ladle dates back to the 1930s when the Perrin process was developed for ladle dephosphorization or slag-aided deoxidation. In modern times the use of synthetic slags has assumed importance because of the requirements for very clean (less than 15 ppm aluminium present as Al_2O_3) and ultra-low sulphur steels (less than 10 ppm sulphur).

The slag carried over from the converter to the ladle has a high oxygen potential. As a result, even if argon bubbling is performed after adding the deoxidants, the total oxygen content of the liquid steel is hardly affected due to the oxidation of aluminium by the reducible oxides present in the slag. Since synthetic slags have a special chemical composition and contain very low FeO, it is necessary to do slag-free tapping or remove the carried over slag by mechanical means (or by ladle-to-ladle transfer). Argon bubbling can then be done to help the flotation of Al_2O_3 and other reaction products and also to promote the exchange reaction at the slag–metal interface. For example, if the synthetic slag has a high sulphide capacity then sulphur can be transferred from the metal to the slag. Similarly, if the activity of Al_2O_3 in the slag is low (e.g. in calcium aluminate slags), then the Al_2O_3 particles floating up from the liquid steel can be absorbed in the top slag; calcium-aluminate-based slags have been successfully used to deoxidize steel, to reduce total oxygen in steel (i.e. to absorb Al_2O_3) as well as to desulphurize steel [2].

In this section the fundamental considerations based on the thermodynamic and kinetic aspects of the exchange treatment by the addition of calcium-aluminate-type synthetic slags will be discussed.

7.4.1 *Thermodynamic considerations*

A wide range of chemical composition of calcium-aluminate-based slags are used in industry. The main requirements of these slags [7] are: low melting point and high fluidity; low oxygen potential; the capacity to absorb Al_2O_3 particles; and high sulphide capacity (in case desulphurization is desired).

The equilibrium phase diagram of the $CaO-Al_2O_3-SiO_2$ system, adapted from Okohira *et al.* [8], is shown in Figure 7.4. If the melting point is to be lower than

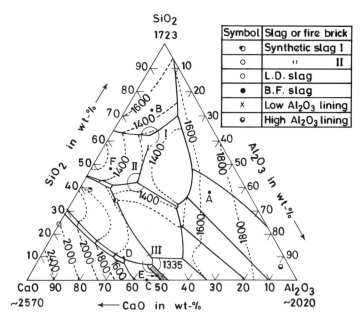

Figure 7.4 Equilibrium phase diagram of $CaO-Al_2O_3-SiO_2$ system showing average compositions of LD slag, BF slag, synthetic slag I, synthetic slag II, low Al_2O_3 lining, high Al_2O_3 lining, and lime-saturated slag (good for desulphurization, marked by hatched region close to $CaO-Al_2O_3$ line). (*Source*: Okohira, K., Sato, N. and Mori, H. (1974), *Trans. Iron Steel Inst. Jpn*, **14**, 337–46.)

1350 °C then typical compositions will be

    $CaO-SiO_2-Al_2O_3$ 22:62:16 slag I   marked by I in Figure 7.4

    $CaO-SiO_2-Al_2O_3$ 38:42:20 slag II   marked by II in Figure 7.4

    $CaO-SiO_2-Al_2O_3$ 50:07:43 slag III marked by hatched region in Figure 7.4

The activities of CaO, Al_2O_3 and SiO_2 in $CaO-SiO_2-Al_2O_3$ slags are shown in Figure 7.5. A low oxygen potential is possible only for slags in the neighbourhood of composition III. For example, under industrial conditions at 1600 °C, 0.05 wt % dissolved aluminium will result in 8 ppm oxygen in liquid steel. Under the same conditions with type III calcium aluminate slag (region III in Figure 7.4) the steel will contain 4 ppm oxygen. This is because the residual oxygen in the liquid steel will be in equilibrium with calcium aluminate particles retained in the steel. If the objective of synthetic slag addition is for deoxidation only, then a final slag in the composition range of 50–55% CaO, 4–5% SiO_2 and rest Al_2O_3 is found to be effective.

From thermodynamic considerations, if the slag is to absorb Al_2O_3 particles then we should start the treatment with slag I and end up in the direction of slag III. However, the best practical results are obtained with type III slag because aluminium oxidation by slag III is minimum.

The sulphide capacity of $CaO-Al_2O_3-SiO_2$ slags [7] is shown in Figure 7.6. A well-deoxidized slag and metal are a prerequisite for desulphurization owing to the exchange reaction

$$[S] + (CaO) = (CaS) + [O] \tag{7.3}$$

The equilibrium constant for the above reaction is

$$K = \frac{h_O a_{CaS}}{h_S a_{CaO}} = 2.1 \times 10^{-3} \text{ at } 1800 \text{ K}$$

$$= 3.06 \times 10^{-3} \text{ at } 1900 \text{ K} \tag{7.4}$$

In the presence of aluminium, the reaction is

$$3[S] + 3(CaO) + 2[Al] = (Al_2O_3) + 3(CaS) \tag{7.5}$$

The sulphur partition as a function of oxygen activity and basicity is shown in Figure 7.7. The chemical composition of synthetic slags used for deoxidation and desulphurization is marked by the hatched region in Figure 7.4. Since these slags are saturated in lime, their melting point is high. In order to promote their melting at lower temperatures CaF_2 can be added or the amounts of Al_2O_3 and SiO_2 can be increased – but at the expense of sulphide capacity. Lime-saturated slags have (S)/[S] ratio of approximately 1000 and this ratio decreases monotonically [2] by the addition of Al_2O_3 and SiO_2 according to

$$\frac{(S)}{[S]} = 1260 - 25(Al_2O_3) - 75(SiO_2) \tag{7.6}$$

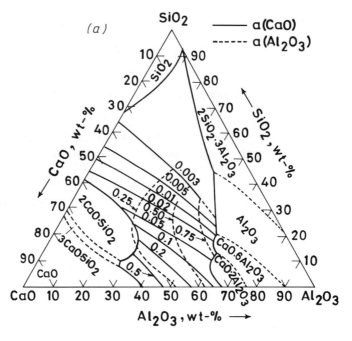

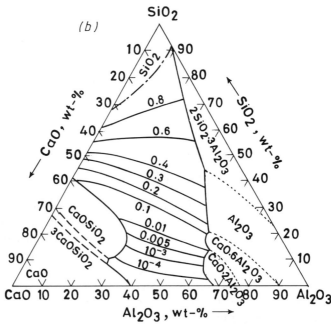

Figure 7.5   Activities in $CaO–Al_2O_3–SiO_2$ system at 1600 °C: (a) CaO and Al_2O_3 activity; (b) SiO_2 activity. (*Source*: Richardson, F. D. (1974) *Physical Chemistry of Melts in Metallurgy*, Vol. 1, Academic Press: London, p. 285.)

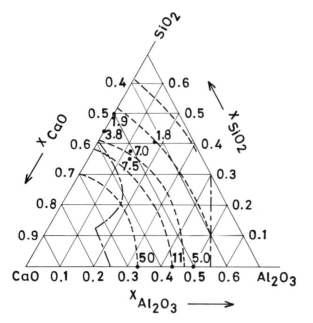

Figure 7.6 Sulphide capacities ($\times 10^4$) at 1650 °C in CaO–Al$_2$O$_3$–SiO$_2$ melts: dashed lines show isocapacity contours; heavy dashed lines show liquid composition limits at 1650 °C. (*Source*: Richardson, F. D. (1974) *Physical Chemistry of Melts in Metallurgy*, Vol. 2, Academic Press: London, p. 296.)

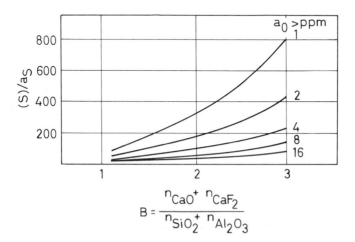

$$B = \frac{n_{CaO} + n_{CaF_2}}{n_{SiO_2} + n_{Al_2O_3}}$$

Figure 7.7 Sulphur partition as a function of oxygen activity and basicity. (*Source*: Lange, K. W. (1988), *Int. Mater. Rev.*, **33**, 53–89.)

For desulphurization, the final slag composition after the assimilation of the deoxidation products should be 60% CaO, 35% Al_2O_3 and 5% SiO_2. The relative amounts of CaO and calcium aluminate in the slag mixture depend upon the deoxidation practice, the initial sulphur content of the metal and the required sulphur reduction.

7.4.2 *Kinetic considerations*

The reaction sites for the oxidation of dissolved aluminium are shown in Figure 7.8. It has been found that in the case of aluminium-killed steel in contact with calcium-aluminate-based synthetic slags, the oxygen content of the metal attains a reasonably low and steady value (ranging from 3 to 5 ppm, depending upon the soluble aluminium content of steel) within 4–8 minutes. The kinetics of the slag–metal reaction after a lapse of 4–8 minutes will therefore be controlled either by the mass transport of aluminium in liquid steel to the slag–metal or refractory–metal interface (i.e. metal phase control, Figure 7.9a), or by the mass transport of oxygen in the slag (i.e. slag phase control, Figure 7.9b), or by both processes simultaneously (i.e. mixed mass transport control, Figure 7.9c).

The treatment for mixed mass transport control is given below.

The equation for the oxidation rate of aluminium due to mass transfer in the metal phase can be written as

$$-\frac{d[\% Al]}{dt} = k_{Al} \frac{A}{V} [\% Al] \tag{7.7}$$

where, as before, A is the nominal interface area (refractory wall–metal interface plus top slag–metal interface), V is the volume of metal and k_{Al} is the mass transport

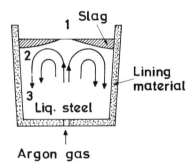

Figure 7.8 Possible sites for oxidation of dissolved aluminium during argon purging in a ladle: 1, aluminium loss caused by air oxidation; 2, aluminium loss caused by FeO and SiO_2 reduction; 3, aluminium loss caused by reaction with refractory lining, skull and dissolved oxygen in steel. (*Source*: Okohira, K., Sato, N. and Mori, H. (1974), *Trans. Iron Steel Inst. Jpn*, **14**, 337–46.)

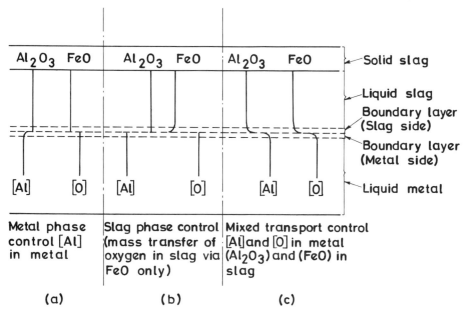

Figure 7.9   Rate-controlling steps for aluminium oxidation: (a) metal phase control;
(b) slag phase control; (c) mixed mass transport control.

coefficient of aluminium in the metal; in an argon-stirred ladle k_{Al} is of the order of
0.0006–0.001 m/s.

The molar flux of FeO from slag ($\dot{n}_{FeO}$ mol/s) can be written as

$$\dot{n}_{FeO} = k_O A(C^b_{FeO} - C^i_{FeO}) \tag{7.8}$$

where k_O is the mass transfer coefficient of oxygen as FeO in the slag phase, A is the
interface area, C^b_{FeO} is the concentration of FeO in the bulk slag and C^i_{FeO} is the
interface concentration. If, for the sake of simplicity, the refractory–metal reaction is
ignored (i.e. in the case of high-alumina lining) and it is assumed that only the FeO
in the top slag oxidizes the aluminium, then from mass balance

$$\frac{V\rho}{100}\frac{48}{54}\frac{d[\% \, Al]}{dt} = 16\dot{n}_{FeO} \tag{7.9}$$

where ρ is the density of liquid steel. In the case of aluminium-killed steel C^i_{FeO} will
be extremely small at the slag–metal interface and hence can be neglected. On
substituting the values of $d[\% \, Al]/dt$ and $\dot{n}_{FeO}$, the critical condition for the
changeover from mass transport control in metal phase to mass transfer control in
slag phase can be written as [8]

$$\frac{V\rho}{100}\frac{48}{54} k_{Al} \frac{A}{V}[\% \, Al] = 16k_O \frac{A}{V} C_{FeO} \tag{7.10}$$

The term C_{FeO} can be converted to wt % FeO by considering the volume and the density of the slag phase. Semi-industrial trials with synthetic slags in a 1 ton ladle [8] have shown that

$$\text{wt \% FeO} = 1.35 \times 10^{-2} \text{ wt \% Al} \tag{7.11}$$

The above equation is represented by the solid line in Figure 7.10; the dashed line represents the limit of mass transfer control in the metal phase only. The region lying between the dashed line and the solid line will thus represent mass transport control by aluminium in the metal whereas the region below the solid line will represent mass transport in the slag phase only. The results of trials [8] have shown that the experimental points lie much below the solid line (shown by the hatched region in Figure 7.10). It can therefore be concluded that in the case of synthetic slags with negligible FeO, the oxidation of aluminium in the metal is predominantly controlled by mass transfer in the slag phase. It will be shown later that during exchange treatment with conditioned slag a similar situation may arise if the top slag, very close to the slag–metal interface, becomes depleted in FeO as a result of the reaction with aluminium dissolved in the metal. The value of k_O in Equation (7.10) (i.e. mass transfer coefficient for the slag phase) is reported to depend strongly upon the FeO content of the slag; a slight lowering of the FeO concentration at the slag–metal interface can drastically lower (by a factor of 1000 or even more) the diffusivity of oxygen in the slag [9, 10]. This can be an asset or a drawback: it becomes an asset if the interface is not disturbed and hence oxygen transfer is hindered due to the low k_O value in an FeO-depleted interface, thereby reducing the oxidation of aluminium; it becomes a drawback if sulphur transfer is to take place from the

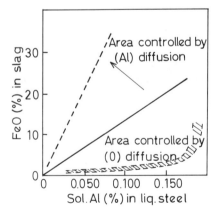

Figure 7.10   Dependence of the kind of reaction control step on FeO content in slag and aluminium content in liquid steel. (*Source*: Okohira, K., Sato, N. and Mori, H. (1974), *Trans. Iron Steel Inst. Jpn*, **14**, 337–46.)

metal to the slag by the exchange reaction

$$[S] + (O^{2-}) = (S^{2-}) + [O] \tag{7.12}$$

High oxygen potential at the interface will inhibit sulphur transfer.

7.4.3 *Some considerations to enhance the kinetics of reactions*

The appropriate chemical compositions of synthetic slags for deoxidation and desulphurization have already been indicated. The melting and proper mixing of slag are as important as the chemical composition so as to enhance the kinetics of transfer from the metal to the slag. Just as in the case of slag conditioning, temperature and composition gradients can develop in the synthetic slag due to heat loss to the ambient atmosphere. It may then be necessary to resort to top slag heating by external means as well as the stirring of metal and slag by argon purging. For a good stirring condition, approximately $1.5-1.8 \, m^3$ (stp)/min of argon gas is required at about 3 m depth in a 200 ton ladle [1]. After vigorous mixing of the slag into the metal, the flow rate of the argon gas should be decreased and the slag should be gently rinsed out of the liquid steel into the top slag. A minimum free-board of approximately 70 cm is required for a 200 ton ladle to avoid overflow. Premelted slags made in an electric furnace or a plasma furnace are better than fused mixtures (so as to avoid hydration and hence hydrogen pickup).

As an alternative to top slag heating, an exothermic mixture of 50% burnt lime, 30% hematite and 12% aluminium powder can be used [1]. Such mixtures are commercially available. Depending upon the sulphur content of the metal to start with and the final sulphur content desired, $5-15 \, kg/ton$ of exothermic mixtures may be added during furnace tapping. For example, by using 10 kg/ton of the exothermic mixture, the steel containing 0.015 to 0.020 wt % sulphur at tapping can be desulphurized to about 0.007 and 0.01 wt % sulphur, respectively. Subsequent hard stirring with argon lowers the sulphur to about 0.003 wt % or less. It may be noted that there is little desulphurization possible at the time of tapping due to high oxygen potential in the liquid steel and slag.

As regards the total oxygen content of steel, industrial data from various plants suggest that for aluminium-killed steel the critical lower limit for soluble aluminium content is 0.03 wt %; if the amount of soluble aluminium is less than 0.03% then total oxygen increases rapidly. In order to minimize oxygen pickup, the top slag should contain the minimum possible FeO, SiO_2 and MnO and the metal surface should not open up (open eye) during vigorous argon stirring. The pickup of oxygen is always more than that calculated on the basis of nitrogen pickup by steel (due to the surface active (or surfactive) nature of oxygen).

Towards the end of the stirring treatment, the argon flow rate is decreased because the presence of undisturbed slag helps to eliminate large (above 100 μm) oxidic particles that can later on generate inclusions in the cast product.

7.5 Exchange treatment with conditioned slag

Exchange treatment with synthetic slag is expensive and time consuming and hence it is recommended usually for special grades of steel. The modern trend is to do slag conditioning. The objective of slag conditioning is twofold:

1. To reduce the loss of soluble aluminium due to reaction with the top slag which is often rich in FeO.
2. To reduce the amount of Al_2O_3 thus produced and hence improve the cleanliness of the cast product.

A mixture of 75% limestone and 25% aluminium when added to slag led to a better recovery of aluminium and a decrease in defects caused by Al_2O_3 (usually called sliver rejects) [3]. Pure lime has also been added at some plants to lower the FeO activity in the slag close to the slag–metal interface and capture large Al_2O_3 particles formed as a result of the slag–metal reaction. This ultimately resulted [3] in the reduction of reject level of the final product – presumably due to the fall in the concentration of large inclusions (above 100 μm in size).

In the argon-stirring route, slag conditioning can be done by adding aluminium pellets on top of the ladle slag just before the start of the stirring treatment. This treatment is, however, found to be not very effective (Hoogovens' experience) due to the fact that slag is already partially deoxidized during tapping (see Table 7.1) itself. Also, during the stirring treatment there is improper mixing of slag with the aluminium added; most of this aluminium somehow finds its way into the liquid steel.

Slag conditioning in the RH-OB route is done just before deoxidation at the RH-OB station. Since slag conditioning has been found to be more effective in the RH-OB route than in the argon-stirring route (Hoogovens' experience), the fundamental considerations based on the thermodynamic and kinetic analysis of slag conditioning in RH-OB will be discussed in this section.

7.5.1 *Thermodynamic considerations*

The equilibrium composition of the metal and slag can be calculated by solving simultaneous equations incorporating equilibrium constants for the elementary reactions and mass balance equations for a closed system (see Chapter 2). Let the liquid steel contain aluminium, manganese, silicon and oxygen in dilute solution. Ladle slag contains reducible oxides including FeO, MnO and SiO_2; CaO is assumed not to interact with the steel. The contribution of P_2O_5 is ignored. The reactions are

$$(FeO) \rightarrow \{Fe\} + [O] \tag{7.13}$$

$$(MnO) \rightarrow [Mn] + [O] \tag{7.14}$$

$$2[Al] + 3[O] \rightarrow (Al_2O_3) \tag{7.15}$$

$$(SiO_2) \rightarrow [Si] + 2[O] \tag{7.16}$$

The equations governing the thermodynamic equilibria can be written as

$$K(\text{FeO}) = \frac{h_\text{O}}{a_\text{FeO}} \tag{7.17}$$

$$K(\text{MnO}) = \frac{h_\text{Mn} h_\text{O}}{a_\text{MnO}} \tag{7.18}$$

$$K(\text{Al}_2\text{O}_3) = \frac{h_\text{O}^3 h_\text{Al}^2}{a_{\text{Al}_2\text{O}_3}} \tag{7.19}$$

$$K(\text{SiO}_2) = \frac{h_\text{O}^2 h_\text{Si}}{a_{\text{SiO}_2}} \tag{7.20}$$

where K is the equilibrium constant, h is the Henrian activity and a_i is the Raoultian activity in the slag phase.

The mass balance equations (in moles) are:

Loss from metal = gain in slag

$$\text{Fe}_0 - \text{Fe} = \text{FeO} - \text{FeO}_0 \tag{7.21}$$

$$\text{Mn}_0 - \text{Mn} = \text{MnO} - \text{MnO}_0 \tag{7.22}$$

$$\text{Al}_0 - \text{Al} = 2(\text{Al}_2\text{O}_3 - (\text{Al}_2\text{O}_3)_0) \tag{7.23}$$

$$\text{Si}_0 - \text{Si} = \text{SiO}_2 - (\text{SiO}_2)_0 \tag{7.24}$$

$$\text{O}_0 - \text{O} = (\text{FeO} - \text{FeO}_0) + 3(\text{Al}_2\text{O}_3 - (\text{Al}_2\text{O}_3)_0) \tag{7.25}$$

where the initial values are indicated by the subscript 0. All initial values are known. The nine end values of FeO, MnO, SiO_2, Al_2O_3, Fe, Mn, Si, Al and O can be obtained by solving the nine non-linear equations, Equations (7.17)–(7.25), on a computer.

The value of K in Equations (7.17)–(7.20) can be obtained from the Gibbs' free energy of formation given in Gaye and Coulombet [11] and Rao [12]. The Henrian activity, h_i, can be written as

$$h_i = f_i \,(\text{wt \% } i) \tag{7.26}$$

The activity coefficients f_i can be calculated from

$$\lg f_i = \sum e_i^j \,(\text{wt \% } j) \tag{7.27}$$

Second-order interaction coefficients are neglected. The first-order interaction coefficients e_i^j can be taken from Sigworth and Elliott [13] (or Table 2.2).

The Raoultian activities in the slag (a_i) are given by

$$a_i = \gamma_i X_i \tag{7.28}$$

where γ_i are the activity coefficients.

Data concerning the activities a_i of oxides in slags are available [14]. However, these data, with a few exceptions, are limited to two- and three-component systems (for instance, the system Al_2O_3–CaO–SiO_2). The activities a_i (or activity coefficients γ_i) in a multicomponent system have been calculated in this work with a computerized version of the Kapoor–Frohberg cell model of the slag [15, 16].

Actual data

A typical analysis of the steel and slag samples at the end of treatment (ET) and at the turret (TU), both with slag conditioning by adding 1000 kg lime in a 300 ton ladle at the RH-OB station before deoxidation, and without slag conditioning, is given in Table 7.3. The steel and slag analyses obtained at the end of treatment (ET) are taken as starting values for the equilibrium calculations in the thermodynamic model described above.

It can be seen from Table 7.3 that the non-conditioned ladle slag of titanium-stabilized low-carbon steel charges (Ti–SulC steel) contains 17.1 wt % FeO and 5.4 wt % MnO at the end of treatment. The conditioned slag contains 13.7 wt % FeO and 6.8 wt % MnO. The percentage of CaO in the conditioned slag increased by 6 wt %. There is a slight reduction in the loss of soluble aluminium between the end of treatment and the turret for the conditioned heat as compared with the non-conditioned heat. The overall results also revealed a positive influence of slag conditioning because the rejects decreased as a result of the increased efficiency of the slag to absorb large Al_2O_3 inclusions.

Table 7.3 Steel and slag compositions for the route via the vacuum degasser: ET = End of Treatment; TU = TUrret; CA = CAlculations. The steel and slag compositions at the end of treatment are used as input for the model calculations.

Slag conditioning Analysis	(wt %)	ET	TU	CA	ET	TU	CA
					Slag conditioning 1000 kg lime at vacuum degasser		
Steel	Fe	99.8	99.8	99.86	99.7	99.7	99.76
	Mn	0.152	0.152	0.171	0.240	0.242	0.265
	Al	0.047	0.043	0.002	0.048	0.047	0.005
	Si	0.004	0.005	0.02	0.005	0.005	0.02
	O	0.000 28	0.000 30	0.0010	0.000 27	0.000 28	0.000 51
Slag	FeO	17.1	14.7	0.205	13.7	12.5	0.082
(1500 kg)	MnO	5.4	4.2	0.135	6.8	6.4	0.06
	Al_2O_3	26.1	25.9	44.0	23.5	24.7	40.7
	SiO_2	10.3	21.9	3.3	8.9	9.8	2.1
	CaO	41.1	33.3	–	47.1	46.6	–

Results

In the thermodynamic equilibrium calculations the total amount of liquid slag is assumed, a priori, to be 1500 kg. The computed results (CA) (included in Table 7.3) show that dissolved aluminium should change from 0.047% to 0.002% for the non-conditioned heat and from 0.048% to 0.005% for the conditioned heat. The actual slag composition is considerably different from the calculated equilibrium values. For example, instead of only a small decrease in FeO and MnO (as actually observed), calculated values show that the final FeO and MnO content should be less than 0.5%. Instead of a calculated reduction in the amount of SiO_2 in the slag from about 10% to about 3% (as predicted by the calculations), an increase is observed. The discrepancy between the calculated values (assuming 1500 kg of liquid reactive slag) and the actual values at the turret may be ascribed to several reasons:

- The actual amount of active liquid slag is smaller than assumed.
- There are strong temperature gradients due to heat loss from the top surface of the slag also and strong concentration gradients within the slag due to incomplete mixing; the slag sample collected on the shop floor may thus not be representative of the bulk slag.
- The transport properties of the slag are such that thermodynamic equilibrium is not established in the available time (between the sampling at the end of treatment and at the turret).
- The composition of slag close to the slag–metal interface is entirely different from the bulk slag, i.e. there is a layer formation (depleted in FeO due to the reaction with aluminium in the bath) close to the slag–metal interface.

It is extremely difficult to account for the above factors in a mathematical model.

The effect of the amount of liquid slag

Model calculations were repeated, taking only 25% of the previously assumed 1500 kg of liquid slag, i.e. 375 kg. The calculated results showed a reduced loss of soluble aluminium (0.006 wt %, which is closer to the actual value as shown in Table 7.3) compared with the previous decrease (0.045 wt % aluminium loss from the metal) predicted from calculations for 1500 kg of liquid slag. The results concerning the FeO, SiO_2 and MnO content of the slag, however, showed no improvement.

The effect of heat losses

It is observed that, soon after tapping, the top layer of the slag begins to solidify due to heat losses to the ambient atmosphere. Unless the bath is turbulent, the mixing in the bulk liquid slag is negligible and therefore sharp temperature and concentration gradients can develop in the slag close to the slag–metal interface.

The fraction of liquid slag in the ladle can be estimated from heat transfer calculations. For a specific case it has been shown [17] that after 30 minutes 2.5 cm of a slag layer (total thickness 5 cm with an initial homogeneous temperature of 1650 °C) still had a temperature of over 1400 °C. However, even an assumption of 375 kg of liquid slag does not bring the predicted results closer to the actual results.

The discrepancies between the calculated and actual results clearly indicate that either the kinetics of reaction is slow or the mixing of slag is incomplete and/or the bulk and interface slag compositions are entirely different. The effect of the transport properties of the slag on the kinetics of the reactions at the slag–metal interface should be considered to explain slow kinetics.

7.5.2   *Kinetic considerations*

The oxygen present in the slag has to diffuse to the slag–metal interface so that the equilibrium between the slag and the metal is eventually established. The oxygen transport through the slag has already been shown to be a rate-limiting factor in the case of synthetic slags low in FeO [8]. Also, it is reported that the oxygen diffusion rate can be influenced enormously by the FeO content [9, 10]; for example, the addition of 0.2 wt % Fe_2O_3 to a slag containing 40 wt % CaO–40 wt % SiO_2–20 wt % Al_2O_3 raises the oxygen diffusion rate by a factor of 10.

As a result of the reaction between the soluble aluminium and the reducible oxides in the slag, an interface layer with strong FeO and Al_2O_3 gradients can form (Figure 7.9c). This interface layer, with its low FeO content, can act as a barrier to the transport of oxygen from the slag to the slag–metal interface. As long as there is incomplete mixing of the slag, the interface layer can remain intact and the kinetics of the reaction between the aluminium in the metal and oxygen ions in the slag will be determined by the oxygen transport through the interface slag layer (depleted in FeO). In such a case, even if a large amount of liquid slag is present, complete equilibrium between the slag and metal will not be established within the available time in spite of the high FeO content of the remaining bulk slag away from the slag–metal interface. This is confirmed by the results obtained in practice at Hoogovens because at the time of sampling at the turret only a small part of the slag near the interface was found to have reacted with the metal and the composition of the bulk slag away from the slag–metal interface did not change very much. It could also explain the low loss of soluble aluminium between the end of treatment and the turret.

Owing to an increase in turbulence in the ladle, the thin interface (or boundary layer which is poor in FeO) can break up leading to the oxidation of soluble aluminium by the FeO in the bulk slag (because the bulk slag has a high oxygen potential). In practice a high loss of soluble aluminium is generally observed between the turret and the tundish and also towards the end of the teeming operation.

Example 7.1 **The kinetics of oxidation of aluminium by top slag during argon stirring in a ladle**

Let the reducible oxides in slag be essentially FeO and MnO. The liquid metal contains dissolved aluminium, manganese and oxygen. The mass transfer coefficients in the slag phase are k_{FeO}, k_{MnO} and $k_{Al_2O_3}$, and in the metal phase are k_O, k_{Mn} and k_{Al}. Mixed transport control of all species is considered. The concentrations at the interface are indicated by superscript i and those in the bulk by superscript b. Individual equilibria at the interface can be written as

$$\{Fe\} + [O] = (FeO)$$

For the sake of simplicity, if ideal behaviour is assumed, then

$$K_{FeO} = \alpha_{FeO} \frac{C^i_{FeO}}{C^i_{Fe} C^i_O} \tag{A}$$

$$[Mn] + [O] = (MnO)$$

$$K_{MnO} = \alpha_{MnO} \frac{C^i_{MnO}}{C^i_{Mn} C^i_O} \tag{B}$$

$$2[Al] + 3[O] = (Al_2O_3)$$

$$K_{Al_2O_3} = \alpha_{Al_2O_3} \frac{C^i_{Al_2O_3}}{(C^i_{Al})^2(C^i_O)^3} \tag{C}$$

where α is the factor used to convert concentration to mole fraction. For example, let the initial total moles of slag be TSLAG and those of metal be TMETAL. Let the initial volume of slag and metal be VSLAG and VMETAL, respectively. The concentrations of FeO, iron and oxygen in Equation (A) are related to mole fractions as follows:

$$X^i_{FeO} = \frac{C^i_{FeO} \, VSLAG}{TSLAG}$$

$$X^i_O = \frac{C^i_O \, VMETAL}{TMETAL}$$

$$X^i_{Fe} = \frac{C^i_{Fe} \, VMETAL}{TMETAL}$$

(For the sake of simplicity $X_{Fe} \approx 1$ and thus C^i_{Fe} is known a priori.) Hence

$$\alpha_{FeO} = \left(\frac{VSLAG}{TSLAG}\right)\left(\frac{TMETAL}{VMETAL}\right)^2$$

Similarly,

$$\alpha_{MnO} = \left(\frac{VSLAG}{TSLAG}\right)\left(\frac{TMETAL}{VMETAL}\right)^2$$

$$\alpha_{Al_2O_3} = \left(\frac{VSLAG}{TSLAG}\right)\left(\frac{TMETAL}{VMETAL}\right)^5$$

For more accurate calculations, the interaction parameters can be incorporated into Equations (A)–(C).

The flux (J) balance equations can be written as

$$J_{Al} = 2J_{Al_2O_3}$$

or

$$k_{Al}(C^b_{Al} - C^i_{Al}) = 2k_{Al_2O_3}(C^i_{Al_2O_3} - C^b_{Al_2O_3}) \tag{D}$$

$$J_{Mn} = J_{MnO}$$

or

$$k_{Mn}(C^b_{Mn} - C^i_{Mn}) = k_{MnO}(C^i_{MnO} - C^b_{MnO}) \tag{E}$$

$$J_O = 3J_{Al_2O_3} + J_{MnO} + J_{FeO}$$

or

$$k_O(C^b_O - C^i_O) = 3k_{Al_2O_3}(C^i_{Al_2O_3} - C^b_{Al_2O_3})$$
$$+ k_{MnO}(C^i_{MnO} - C^b_{MnO})$$
$$+ k_{FeO}(C^i_{FeO} - C^b_{FeO}) \tag{F}$$

The equations (A)–(F) constitute six equations with six unknowns, namely C^i_O, C^i_{Mn}, C^i_{Al}, $C^i_{Al_2O_3}$, C^i_{MnO} and C^i_{FeO}. These equations can be solved iteratively to compute the interfacial concentrations. During a given time step, Δt, the individual fluxes J_{Al}, J_{Mn}, J_O, $J_{Al_2O_3}$, J_{MnO}, J_{FeO} can be calculated to give new bulk slag and metal concentrations for the next time step. For example,

$$J_{Al} = k_{Al}(C^b_{Al} - C^i_{Al}) \quad \text{or} \quad -\frac{\Delta Al}{\Delta t} = \frac{A}{V} k_{Al}(C^b_{Al} - C^i_{Al})$$

where A is the area of the slag–metal interface (m^2) and V is the volume of metal (m^3). In this way the new bulk concentrations of aluminium, manganese and oxygen in the metal and of Al$_2$O$_3$, FeO and MnO in the slag can be calculated as a function of time.

Typical values of the mass transfer coefficients in a gas-stirred bath may be taken as

$$k_{Al} = k_O = k_{Mn} = 5 \times 10^{-4} \, \text{m/s}$$

$$k_{Al_2O_3} = k_{FeO} = k_{MnO} = 8 \times 10^{-5} \, \text{m/s}$$

The values of VMETAL, TMETAL and VSLAG, TSLAG can be computed from the known initial composition and mass of the metal and slag.

7.6 Deoxidation and decarburization under reduced pressure of carbon monoxide gas

Towards the end of primary steelmaking, the carbon and oxygen contents of liquid steel are slightly in excess of equilibrium concentrations. If this steel is subjected to stirring treatment with argon gas in the oxygen steelmaking converter then the reaction between dissolved carbon and dissolved oxygen, $[C] + [O] = CO(g)$, is favoured owing to the low partial pressure of carbon monoxide gas in the rising gas bubbles. Hence, as a result of argon stirring in the furnace (i.e. in a combined-blown converter or in an arc furnace) both the carbon and oxygen contents of liquid steel are lowered and the FeO content of the slag in the converter is reduced by 1–2%. The equilibrium carbon and oxygen contents of steel at different reduced pressures of carbon monoxide gas are shown in Figure 2.10. It is possible to reduce the pressure of carbon monoxide gas in liquid steel by purging with argon gas and/or by vacuum treatment in RH, RH-OB, etc.

The dissolved carbon can also react with the combined oxygen present in the refractory lining material of a ladle (i.e. oxygen present in SiO_2, FeO, etc.). The thermodynamic feasibility of the carbon–refractory reaction at different pressures of carbon monoxide gas is shown in Figure 7.11 [18].

A detailed theoretical analysis of the various possible rate-controlling steps in the carbon–oxygen reaction has been described in the literature [19, 20]. The treatment given below, except for the cases of mass transport control, is adapted from Suzuki and Mori [19].

The overall reaction of carbon monoxide formation can be broken down into several steps and any one or a combination of these steps can become rate controlling.

1. Transport of dissolved carbon from bulk metal to the metal–argon bubble interface:
The flux of carbon can be written as

$$J_C = J_{CO} = -\frac{V}{A}\left(\frac{dC_C^b}{dt}\right) = k_C(C_C^b - C_C^i) \tag{7.29}$$

where J_C (kmol/m^2 s) is the flux of carbon, J_{CO} is the flux of carbon monoxide, V is the volume of metal (m^3), A is the area of the bubble–metal interface (m^2), C_C^b is the concentration of carbon in bulk metal (kmol/m^3), C_C^i is the concentration of carbon at the bubble–metal interface and k_C is the mass transfer coefficient of carbon in the metal.

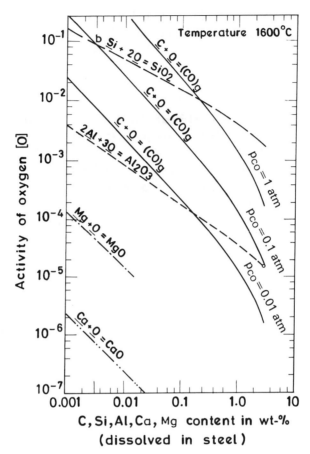

Figure 7.11   Equilibrium oxygen activity in liquid steel for different dissolved contents of carbon, silicon, aluminium, calcium and magnesium (wt%). (*Source*: Kreutzer, H. W. (1972), *Stahl und Eisen*, **92**, 716–24.)

2. Transport of dissolved oxygen from bulk metal to the bubble–metal interface:
Using the notation similar to that in Equation (7.29):

$$J_O = J_{CO} = -\frac{V}{A}\left(\frac{dC_O^b}{dt}\right) = k_O(C_O^b - C_O^i) \tag{7.30}$$

where the subscript O refers to oxygen.

3. Chemical reaction between carbon and oxygen at the bubble–metal interface:
It can be easily shown that the flux of carbon monoxide gas is given by

$$J_{CO} = k_r^o(C_C^i C_O^i - p_{CO}^i/m) \tag{7.31}$$

where k_r^o is the reaction rate constant, p_{CO}^i is the partial pressure of carbon monoxide gas at the bubble–metal interface and m (m^6 atm/mol^2) is the equilibrium constant (in terms of molar concentration):

$$m = \frac{p_{CO}^e}{C_C^e C_O^e} \tag{7.32}$$

where the superscript e refers to equilibrium value.

4. Transport of carbon monoxide gas from the bubble–metal interface into the bulk gas: The flux of carbon monoxide gas is given by

$$J_{CO} = \frac{k_g}{RT_f}(p_{CO}^i - p_{CO}^b) \tag{7.33}$$

where k_g is the mass transfer coefficient of carbon monoxide gas, R is the gas constant and T_f is the gas-side film temperature.

On eliminating p_{CO}^i and p_{CO}^b with the help of Equations (7.31) and (7.32):

$$J_{CO} = k_r(C_C^i C_O^i - C_C^e C_O^e) \tag{7.34}$$

where

$$k_r = \frac{1}{k_r^o} + \frac{RT_f}{mk_g} \tag{7.35}$$

Similarly, eliminating C_C^i and C_O^i in Equation (7.34) with the help of Equations (7.29) and (7.30):

$$J_{CO} = \frac{k_C k_O}{2}\left(\frac{1}{k_r} + \frac{C_C^b}{k_O} + \frac{C_O^b}{k_C}\right)$$
$$\times \left[1 - \left(1 - \frac{4(C_C^b C_O^b/k_C k_O)(1 - C_C^e C_O^e/C_C^b C_O^b)}{(1/k_r + C_C^b/k_O + C_O^b/k_C)^2}\right)^{1/2}\right] \tag{7.36}$$

In the argon bubble or in the ambient atmosphere, during vacuum treatment, $p_{CO} \approx 0$ and hence $C_C^e C_O^e/C_C^b C_O^b \ll 1$. With this simplification Equation (7.36) can be written as

$$J_{CO} \simeq \frac{k_C k_O}{2}\left(\frac{1}{k_r} + \frac{C_C^b}{k_O} + \frac{C_O^b}{k_C}\right)$$
$$\times \left[1 - \left(1 - \frac{(4C_C^b C_O^b/k_C k_O)}{(1/k_r + C_C^b/k_O + C_O^b/k_C)^2}\right)^{1/2}\right] \tag{7.37}$$

This equation describes a general case of mixed control, i.e. chemical reaction plus oxygen diffusion plus carbon diffusion.

The square root in Equation (7.37) can be simplified by considering only the first term in a binomial expansion (i.e. $(1 - x)^n = (1 - nx)$) provided that

$$4C_C^b C_O^b / k_C k_O \ll \left(\frac{1}{k_r} + \frac{C_C^b}{k_O} + \frac{C_O^b}{k_C} \right)^2 \tag{7.38}$$

Hence

$$J_{CO} \simeq \frac{C_C^b C_O^b}{1/k_r + C_C^b/k_O + C_O^b/k_C} \tag{7.39}$$

Several specific cases can now be considered.

- **Pure chemical reaction control:**   The limiting condition becomes

$$\frac{1}{k_r} \gg \frac{C_C^b}{k_O} + \frac{C_O^b}{k_C} \tag{7.40}$$

and hence from Equation (7.39):

$$J_{CO} = k_r C_C^b C_O^b \tag{7.41}$$

- **Chemical reaction plus carbon diffusion control (or mixed control):**   The limiting condition becomes

$$\frac{1}{k_r} + \frac{C_O^b}{k_C} \gg \frac{C_C^b}{k_O} \tag{7.42}$$

Hence from Equation (7.39)

$$J_{CO} \simeq \frac{C_C^b C_O^b}{1/k_r + C_O^b/k_C} \tag{7.43}$$

- **Carbon diffusion control only:**   The limiting condition becomes

$$\frac{C_O^b}{k_C} \gg \frac{1}{k_r} + \frac{C_C^b}{k_O} \tag{7.44}$$

and from Equation (7.39)

$$J_{CO} = k_C C_C^b \tag{7.45}$$

- **Oxygen diffusion control only:**   Following the same procedure as for carbon diffusion control:

$$J_{CO} = k_O C_O^b \tag{7.46}$$

- **Mixed mass transport control of carbon and oxygen in metal phase only:**   From Equations (7.29) and (7.30), for the case of mixed mass transport control in the metal phase

$$J_O = J_C$$

or

$$k_O(C_O^b - C_O^i) = k_C(C_C^b - C_C^i) \tag{7.47}$$

Also, assuming thermodynamic equilibrium at the interface, from Equation (7.32)

$$m = \frac{p_{CO}}{[C_C^i][C_O^i]} \tag{7.48}$$

The equations (7.47) and (7.48) constitute two simultaneous equations in two unknowns and can be solved iteratively, for instance by the Newton–Raphson method.

● **Mixed mass transport control of carbon and oxygen in metal phase and of carbon monoxide in gas phase:**   For this case Equation (7.48) becomes

$$m = \frac{p_{CO}^i}{C_C^i C_O^i} \tag{7.49}$$

Since $J_{CO} = J_O$ (or $J_{CO} = J_C$), from Equations (7.30) and (7.33) we get

$$k_O(C_O^b - C_O^i) = \frac{k_g}{RT_f}(p_{CO}^i - p_{CO}^b) \tag{7.50}$$

The Equations (7.47), (7.49) and (7.50) constitute three simultaneous equations with three unknowns (C_C^i, C_O^i and p_{CO}^i). These simultaneous equations can be solved iteratively.

The specific cases discussed have been tested by measuring the concentration of carbon and oxygen as a function of time in an argon-stirred bath [19]. The experimental data show the following:

● For carbon greater than 0.03 wt %, the rate is controlled by oxygen mass transfer in the metal only; the value of k_O is approximately 3.5×10^{-4} m/s.
● For oxygen greater than 0.06 wt %, the rate is controlled by mass transfer of carbon only and the value of k_C is approximately 5.1×10^{-4} m/s.
● The changeover from oxygen mass transfer control to carbon mass transfer control occurs at a ratio of $k_O/k_C = 3.5 \times 10^{-4}/5.1 \times 10^{-4} = 0.69$, as shown in Figure 7.12 [19].
● Mixed mass transport control and mixed control (chemical reaction plus mass transfer), when assumed as the rate-controlling step, do not fit with the observed results.

The condition that k_O/k_C determines the changeover from oxygen mass transport control to carbon mass transfer control can be derived from Equations (7.29) and (7.30). For example, if the rate is controlled by the mass

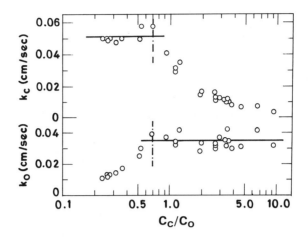

Figure 7.12   Plot of k_O and k_C as a function of C_C/C_O; C_C is concentration of dissolved carbon and C_O is concentration of dissolved oxygen in steel. (*Source*: Suzuki, K. and Mori, K. (1977), *Trans. Iron Steel Inst. Jpn*, **17**, 136–42.)

transfer of carbon

$$k_O(C_O^b - C_O^i) > k_C(C_C^b - C_C^i) \tag{7.51}$$

Since the reaction at the bubble–metal interface occurs instantaneously as a result of the low partial pressure of carbon monoxide gas in the bubble, we can neglect C_C^i and C_O^i and hence

$$\frac{C_C^b}{C_O^b} < \frac{k_O}{k_C} \tag{7.52}$$

($=0.69$ as mentioned above).

The experimental data in Figure 7.12 also show that for $C_C/C_O < 0.69$, the k_C value is constant (i.e. the reaction is controlled by the mass transport of carbon) and for $C_C/C_O > 0.69$, the k_O value is constant. The overall kinetics of decarburization during argon stirring, depending upon the rate-controlling step, will thus be determined by the initial carbon (or oxygen) content and the mass transfer coefficient (k_O or k_C). The effect of the intensity of argon stirring on enhancing the mass transfer coefficient has been discussed in Chapters 1 and 3 and also in section 7.3. In practice, even the location and design of gas injecting nozzles (or porous plugs) fitted at the bottom of the vessel are known to influence the final $[C] \times [O]$ product.

The kinetics of decarburization during vacuum treatment (with or without argon stirring) is influenced by the degree of vacuum, the initial carbon and oxygen content, and the circulation rate of the metal. It is clear from region A of Figure 7.13 that if carbon and oxygen in the bath are 0.04 and 0.05 wt %, respectively, at the

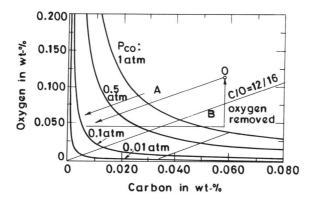

Figure 7.13 Carbon–oxygen equilibria as a function of the partial pressure of carbon monoxide gas; the path of the carbon–oxygen reaction during vacuum treatment is indicated by arrows. (*Source*: Kreutzer, H. W. (1972), *Stahl und Eisen*, **92**, 716–24.)

start then no external source of oxygen will be needed to obtain 0.01 wt % carbon at 0.01 atm pressure of carbon monoxide gas. However, if the starting carbon level is greater than 0.05 wt % (corresponding to region B in Figure 7.13) then the oxygen supply from an external source will be necessary to bring down carbon to 0.01 wt %. The effect of vacuum in a 100 ton RH vessel [21] as well as the effect of the initial carbon and oxygen content of liquid steel on the progress of decarburization and deoxidation reactions are shown in Figure 7.14. The main advantage of vacuum treatment is that the low-carbon steels produced at low pressures of carbon monoxide gas do not require intensive deoxidation. This helps to minimize the formation of deoxidation products like Al_2O_3 and thereby to produce cleaner steels. Treatment in RH-OB (i.e. RH with the additional facility of oxygen injection) helps to produce ultra-low carbon steels.

Several empirical correlations have been proposed to calculate the circulation rate in an RH and RH-OB unit [21, 22]. For a 100 ton RH unit the following correlation has been reported to be satisfactory:

$$Q \text{ (tons/min)} = 1.14G^{1/3}D^{4/3} \left[\ln(p_1/p_2)\right]^{1/3} \tag{7.53}$$

where G is the gas flow rate $(m^3 (\text{stp})/\text{min})$, D is the inner diameter of the snorkel (m), p_1 and p_2 are pressures (atm) inside the bubble at the injection point and at the top of the up leg. On the shop floor, empirical relations are established between the circulation rate and degassing rate, decarburization rate, etc.

In the production of high-alloy steels containing chromium (namely, in the AOD process), the reaction

$$(Cr_3O_4) + 4[C] = 3[Cr] + 4CO(g)$$

$$\lg K' = \lg\left(\frac{(\% \, Cr)^3 (p_{CO})^4}{(h_C)^4}\right) = -\frac{48\,880}{T} + 31.96 \tag{7.54}$$

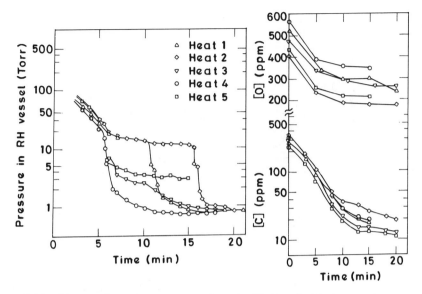

Figure 7.14 Changes in carbon content of steel with time just below up leg; pressure of upper part of vessel; change in dissolved oxygen content of steel with time during treatment in RH vessel. (*Source*: Kuwabara, T., Umezawa, K., Mori, K. and Watanabe, H. (1988), *Trans. Iron Steel Inst. Jpn*, **28**, 305–14.)

is taken into account for the decarburization and recovery of chromium at high temperature and low pressures of carbon monoxide gas.

7.7   Degassing of steel

The thermodynamics of the dissolution of oxygen, nitrogen and hydrogen gases in steel has been discussed in Chapter 2. The solubility of hydrogen and nitrogen in liquid steel at 1600 and 1700 °C is shown in Figure 7.15 [18]. The kinetics of deoxidation and decarburization under reduced pressure of carbon monoxide gas has been discussed in section 7.6.

The kinetics of the degassing of hydrogen and nitrogen can be studied under three different cases:

Case 1: Degassing during carbon boil, i.e. when carbon monoxide gas evolves due to the reaction between dissolved carbon and oxygen; carbon monoxide bubbles provide local sites of low partial pressure of hydrogen and nitrogen.

Case 2: Degassing during argon purging; argon bubbles provide local sites of low partial pressure of hydrogen and nitrogen.

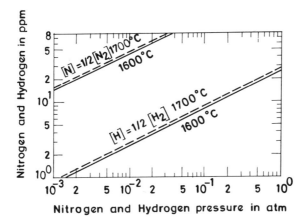

Figure 7.15 Solubility of hydrogen and nitrogen at 1600 and 1700 °C. (*Source:* Kreutzer, H. W. (1972), *Stahl und Eisen*, **92**, 716–24.)

Case 3: Degassing during vacuum treatment; the ambient atmosphere itself has low partial pressure of hydrogen and nitrogen.

A combination of any two or all of the above cases also can exist in actual practice. For example, a ladle can be kept in an evacuated vessel and argon can be purged through the bottom (case 2 + case 3); in the RH process degassing and decarburization occur simultaneously (case 1 + case 2 + case 3).

The adsorption and desorption of hydrogen are reported to be fast. If equilibrium conditions are assumed at the bubble–metal interface, then during carbon boil

$$\frac{d[H]}{dt} = \frac{2}{12} \frac{[H]^2}{K_{H_2}} \frac{d[C]}{dt} \qquad (7.55)$$

where

$$K_{H_2} = \frac{[H]^2}{p_{H_2}}$$

(Henrian behaviour of [O], [C] and [H] has been assumed).

Similarly, during argon purging, if the volumetric flow rate of argon, V_{Ar} (m³(stp)/min), is much greater than that of hydrogen removal, V_{H_2} (m³(stp)/min), then

$$V_{H_2} = -112\left(\frac{d[H]}{dt}\right)W = V_{Ar}\frac{[H]^2}{K_{H_2}} \qquad (7.56)$$

where W is the weight of bath in tons. For a known flow rate of argon, the hydrogen removal rate can be calculated.

Various models for nitrogen adsorption/desorption have been discussed in Chapter 3 (see also Appendix B). In the absence of surfactive elements and during carbon boil

$$p_{N_2} = \frac{1}{28}\frac{d[N]}{dt} \bigg/ \left(\frac{1}{28}\frac{d[N]}{dt} + \frac{1}{12}\frac{d[C]}{dt}\right) = \frac{[N]^2}{K_{N_2}} \tag{7.57}$$

where $K_{N_2} = [N]^2/p_{N_2}$ (Henrian behaviour of nitrogen is assumed).
Assuming that $d[N]/dt \ll d[C]/dt$,

$$\frac{1}{[N]^2}\frac{d[N]}{dt} = \frac{28}{12}\frac{1}{K_{N_2}}\frac{d[C]}{dt} \tag{7.58}$$

Thus, for a known decarburization rate, the degassing rate of nitrogen can be calculated.

It is found in practice that during carbon boil a considerable amount of nitrogen is removed but there is little or no removal of dissolved nitrogen during argon purging only (i.e. in the absence of the carbon–oxygen reaction).

A simple kinetic model for the simultaneous degassing of hydrogen and nitrogen during argon purging is developed below by assuming the mass transport of elements in the metal phase to be rate limiting.

From mass balance:

$$\dot{n}_{CO} = -\frac{10}{16}\frac{d[O]}{dt} \tag{7.59}$$

$$\dot{n}_{H_2} = -\frac{10}{2}\frac{d[H]}{dt} \tag{7.60}$$

$$\dot{n}_{N_2} = -\frac{10}{28}\frac{d[N]}{dt} \tag{7.61}$$

Let $\dot{n}_{Ar}$ be the flow rate of argon gas. Then the total exit gas flow rate, $\dot{n}_T$ (mol/s), is given by

$$\dot{n}_T = \dot{n}_{CO} + \dot{n}_{N_2} + \dot{n}_{H_2} + \dot{n}_{Ar} \tag{7.62}$$

The partial pressures of gaseous species in the bubble can be related to the interfacial concentration of dissolved oxygen, hydrogen and nitrogen through the equilibrium constant:

$$p_{CO} = \dot{n}_{CO}/\dot{n}_T = K_{CO}[C][O_i] \tag{7.63}$$

(only the mass transfer of oxygen is assumed)

$$p_{H_2} = \frac{\dot{n}_{H_2}}{\dot{n}_T} = \frac{[H_i]^2}{K_{H_2}} \tag{7.64}$$

$$p_{N_2} = \frac{\dot{n}_{N_2}}{\dot{n}_T} = \frac{[N_i]^2}{K_{N_2}} \tag{7.65}$$

where the subscript i refers to the interface and Henrian behaviour of the elements is assumed for the sake of simplicity. The molar rates of evolution of the gases are

$$\dot{n}_{CO} = k_O \frac{AM_T}{VMETAL} ([O] - [O_i]) \tag{7.66}$$

$$\dot{n}_{H_2} = \frac{k_{H_2}}{2} \frac{AM_T}{VMETAL} ([H] - [H_i]) \tag{7.67}$$

$$\dot{n}_{N_2} = \frac{k_{N_2}}{2} \frac{AM_T}{VMETAL} ([N] - [N_i]) \tag{7.68}$$

where k is the mass transfer coefficient (m/s) and M_T is the total moles of metal, VMETAL is the volume of metal (m³) and A is the total metal–gas interface area (m²). The equations (7.59)–(7.65) constitute seven equations with seven unknowns ($[N_i]$, $[O_i]$, $[H_i]$, $\dot{n}_{CO}$, $\dot{n}_{H_2}$, $\dot{n}_{N_2}$ and $\dot{n}_T$) and can be solved by a suitable iterative method to give the gas evolution rates at each time step (see Example 3.3 for the method of calculating bulk concentrations as a function of time). Approximate values of k_H, k_N and k_O at 1600 °C are: $k_H = 2.5 \times 10^{-3}$ m/s and $k_O = k_N = 5 \times 10^{-4}$ m/s. It is, however, difficult to estimate correctly the value of area A. In such cases experimentally determined values of capacity mass transfer coefficient (the product $k \times A$) is used in the kinetic model.

Typical concentrations of hydrogen and nitrogen in liquid steel, at the end of secondary steelmaking, may range from 1 to 3 ppm for hydrogen and from 10 to 40 ppm for nitrogen. Actual values depend upon the pickup and removal of hydrogen and nitrogen at various stages of steelmaking. For example, during the initial blow period in an oxygen steelmaking converter, nitrogen is removed as a result of carbon monoxide gas evolution. During the last stages of the blow, nitrogen pickup can take place, especially during the raising and lowering of the lance. Similarly, if the deoxidized steel is exposed to the ambient atmosphere (e.g. during argon purging) nitrogen pickup can take place at the exposed metal surface (Figure 7.8). Enhanced nitrogen pickup (10–30 ppm) can take place during Ca–Si injection because of the absence of surfactive oxygen and sulphur. Moisture in charge materials raises hydrogen content; lime-saturated $CaO–SiO_2–Al_2O_3$ slags are a source of hydrogen pickup. Moisture from the refractory also raises hydrogen content. The variation in nitrogen content of steel at different stages of the process is shown in Figure 7.16 [23].

In general, oxidizing slags enhance hydrogen pickup whereas reducing slags promote nitrogen pickup. Care has to be taken at each process step in order to prevent hydrogen and nitrogen from getting into the steel [23].

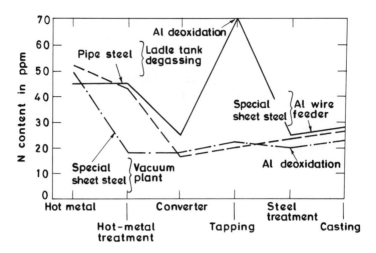

Figure 7.16   Variation of nitrogen content; process routes with different hot metal treatment and aluminium oxidation. (*Source*: Hees, E. (1990).)

Example 7.2 Hydrogen removal by argon purging

(This example is adapted from the work of F. Oeters (1985) 'Kinetic treatment of chemical reactions in emulsion metallurgy', *Steel Res.*, **56**, 69–74.)

Let us assume that the mass transfer of hydrogen in the metal phase is rate controlling. Let [H] be the concentration of hydrogen in metal and C_{H_2} be the concentration of hydrogen in gas. The flux of hydrogen (J_H) can be written as

$$J_H = k_H([H] - K'_H \sqrt{(C_{H_2})})$$

where K'_H is the equilibrium constant based on concentration. Sieverts law constant (K_H) is related to (K'_H) by:

$$K'_H = \frac{[H]}{\sqrt{C_{H_2}}}$$

$$= 10^{-6} K_H \frac{\rho_{Fe}\sqrt{RT}}{M_{H_2}}$$

where M_{H_2} is molecular weight of hydrogen gas and ρ_{Fe} is density of iron.

Expressing J_H in terms of the molar flux density:

$$J_H = \frac{dC_{H_2}}{dt}\left(\frac{V}{A}\right)_B \left(\frac{k\,moles^{1/2}}{m^{3/2}}\right)$$

where the subscript B refers to the bubble with area A and volume V. The ratio $(A/V)_B$ changes with the height (h) of the bubble in the melt. If $(A/V)_B^o$ is the value at the orifice of the nozzle, then

$$\frac{(A/V)_B}{(A/V)_B^o} = \left(\frac{V_B^o}{V_B}\right)^{1/3} = \left(\frac{p_o + g\rho_{Fe}h_t}{p_o + g\rho_{Fe}(h_t - h)}\right)^{-1/3} = (1 - z)^{1/3} \tag{A}$$

where

$$Z = \frac{g\rho_{Fe}h}{p_o + g\rho_{Fe}h_t} \tag{B}$$

ρ_{Fe} is the density of iron, h_t is the height of melt, p_o is the pressure above the melt, g is the acceleration due to gravity and h is vertical position in ladle.

$(A/V)_B^o$ can be determined from the diameter of the bubble. From Equations (A) and (B)

$$J_H = \frac{dC_{H_2}}{dt} \frac{(1 - z)^{1/3}}{(A/V)_B^o} \tag{C}$$

The average rising velocity of bubbles (v_B) is

$$v_B = dh/dt \tag{D}$$

It is assumed that the slip velocity between the bubbles and liquid in the plume zone is nearly zero. The liquid volume flow, $\dot{V}_{LP}$, in the plume zone is given by

$$\dot{V}_{LP} = 5.64 \left[\tan^4\left(\frac{\alpha_o}{2}\right)(h + h_o)^4 \frac{RT}{\rho_g V_N} \ln\left(1 + \frac{g\rho_{Fe}h_t}{p_o}\right) Q_G \right]^{1/3} \tag{E}$$

where

α_o = opening angle of plume zone
ρ_g = density of gas
$V_N = 22.4 \times 10^{-3}$
Q_G = volume flow rate of gas
h_o = distance of vertex of plume zone from metal surface.

The average rising velocity v_B is calculated by assuming a Gaussian distribution of the variation of velocity within the plume zone. From Equations (B) and (D)

$$dt = \frac{p_o + g\rho_{Fe}h_t}{v_B g\rho_{Fe}} dz \tag{F}$$

On substituting dt from Equation (F) into Equation (C):

$$\frac{dC_{H_2}}{([H]/K_H') - \sqrt{(C_{H_2})}} = 2\frac{\sqrt{(C_{H_2})} dC_{H_2}}{([H]/K_H') - \sqrt{C_{H_2}}} \tag{G}$$

Integrating Equation (G) with the boundary conditions $C_{H_2} = 0$ at $z = 0$ (at $h = 0$) and $C_{H_2} = C_{H_2}^z$ at $z = Z$ (at $h = h_t$), we obtain

$$\sqrt{(C_{H_2}^z)} + \left(\frac{[H]}{K'_H}\right)\ln\left(1 - \frac{K'_H\sqrt{(C_{H_2}^z)}}{[H]}\right) = -\tfrac{3}{8}K'_H k_H \left(\frac{A}{V}\right)_B^o \left(\frac{p_o + g\rho_{Fe}h_t}{V_B g\rho_{Fe}}\right)(1-2)^{4/3} \quad \textbf{(H)}$$

The value of $C_{H_2}^z$ can be found by applying the Newton–Raphson method, and for each time step, Δt, one can calculate from the mass balance equations the change in hydrogen concentration $\Delta[H]$:

$$\Delta[H] = -\Delta t V_M C_{H_2}^z Q_G \frac{RT_M}{p_o V_N} \quad \textbf{(I)}$$

where V_M is the molar volume of metal and T_M is the temperature of metal.

References

1. Turkdogan, E. T. (1988) 'Reaction of liquid steel with slag during furnace tapping', *Ironmaking and Steelmaking*, **15**, 311–17.
2. Lange, K. W. (1988) 'Thermodynamic and kinetic aspects of secondary steelmaking processes', *Int. Mater. Rev.*, **33**, 53–89.
3. Bieniosek, T. H., Kupchik, B. S. and Ahlborg, K. C. (1989) 'Ladle slag deoxidation with aluminum–limestone mixtures', *Steelmaking Conf. Proc.*, Vol. 72 1989, ISS-AIME: Warrendale, pp. 409–11.
4. Pluschkell, W., Redenz, B. and Schürmann, E. (1981) 'Kinetics of aluminum oxidation during argon injection into liquid steel', *Arch. Eisenhüttenwes.*, **52**, 85–90.
5. Karl, K. H., Abratis, H., Maas, H. and Wahlster, M. (1974) 'Reaktionen sauerstoffaffiner Begleitelemente des Eisens mit eisenoxidhaltigen basischen Schlacken' *Arch. Eisenhüttenwes.*, **45**, 9–16.
6. Mietz, J. and Brühl, M. (1990) 'Model calculations for mass transfer with mixing in ladle metallurgy', *Steel Res.*, **61**, 105–12.
7. Richardson, F. D. (1974) *Physical Chemistry of Melts in Metallurgy*, Vol. 2, Academic Press: London.
8. Okohira, K., Sato, N. and Mori, H. (1974) 'Liquid steel treatment in the ladle with non-oxidizing synthetic slag', *Trans. Iron Steel Inst. Jpn*, **14**, 337–46.
9. Sasabe, M. and Sato, N. (1974) 'Permeability, diffusivity, and solubility of oxygen gas in liquid slag', *Metall. Trans.*, **5**, 2225.
10. Sasabe, M. and Kinoshita, Y. (1980) 'Permeability of oxygen through molten $CaO–SiO_2–Al_2O_3$, $Fe_2O_3–CaO–SiO_2–Al_2O_3$ and $CaF_2–CaO–SiO_2–Al_2O_3$ systems', *Trans. Iron Steel Inst. Jpn*, **20**, 801–9.
11. Gaye, H. and Coulombet, G. (1985) 'Données thermochimiques et cinétiques relatives à certains matériaux sidérurgiques' *ECSC Report* EUR 9428 FR.
12. Rao, Y. K. (1985) *Stoichiometry and Thermodynamics of Metallurgical Processes*, Cambridge University Press: London.
13. Sigworth, G. K. and Elliott, J. F. (1974) 'The thermodynamics of dilute iron alloys', *Met. Sci.*, **8**, 298–310.

14. *Slag Atlas*, Verlag Stahleisen: Düsseldorf, 1981.

15. Kapoor, M. L. and Frohberg, M. G. (1970) 'The electrolytic dissociation of molten slags and its importance for metallurgical reactions, *Arch. Eisenhüttenwes.*, **41**, 1035–40.

16. Kapoor, M. L. and Frohberg, M. G. (1974) 'The calculation of thermodynamic values and structural properties of liquid binary silicate systems', *Arch. Eisenhüttenwes.*, **45**, 663–9.

17. Szekely, J. and Lee, R. G. (1968) 'The effect of slag thickness on heat loss from ladles holding molten steel', *Trans. Metall. Soc. AIME*, **242**, 961–5.

18. Kreutzer, H. W. (1972) 'Vakuumbehandlung von flüssigen Stahl', *Stahl und Eisen*, **92**, 716–24.

19. Suzuki, K. and Mori, K. (1977) 'Rate of desorption of CO from liquid iron', *Trans. Iron Steel Inst. Jpn*, **17**, 136–42.

20. Ito, K., Amano, K. and Sakao, H. (1977) 'On the kinetics of CO degassing from molten iron by argon flow', *Trans. Iron Steel Inst. Jpn*, **17**, 685–92.

21. Kuwabara, T., Umezawa, K., Mori, K. and Watanabe, H. (1988) 'Investigation of decarburization behavior in RH-reactor and its operation improvement', *Trans. Iron Steel Inst. Jpn*, **28**, 305–14.

22. Sheshadri, V. and De Souza Costa, S. L. (1986) 'Cold model studies of R.H. degassing process', *Trans. Iron Steel Inst. Jpn*, **26**, 133–8.

23. Hees, E. (1990) 'Equipment and processes in basic oxygen steelmaking plant for the adjustment of ultra-low content of C, P, S and N', *Metall. Plant and Techn. Int.*, **13**, 26–37.

8 *Injection metallurgy*

8.1 Injection of powdered reagents into liquid steel

The primary objective of introducing powdered reagents into liquid steel, with the help of either cored wire or a carrier gas, is for desulphurization, deoxidation, alloy dissolution and inclusion shape control. A wide variety of powders, mechanically mixed or premelted and granulated, are used in industry for this purpose:

- $Ca-Si$, CaC_2, Mg, $CaO + CaF_2$, $CaO + Al_2O_3$, $CaO + Al_2O_3 + CaF_2$ for desulphurization.
- $Fe-Si$, $CaCN_2$, graphite, NiO, MoO_2, $Fe-B$, $Fe-Ti$, $Fe-Zr$, $Fe-W$, $Si-Zr$, Pb, $Fe-Se$, Te for alloying.
- $Ca-Si$, Al, $Ca-Si-Ba$, $Ca-Si-Mn$ for deoxidation and inclusion shape control.

Injection technology and equipment (not included in this book) are well advanced and available on a commercial basis. The fundamentals of the slag–metal reaction during powder injection, with special reference to desulphurization treatment, are discussed in section 8.2. Alloying by powder injection is discussed in section 8.3 and inclusion shape control by calcium treatment is discussed in section 8.4.

8.2 Desulphurization during powder injection with carrier gas

The interaction between the injected gas plus powder and the liquid metal is complex because of a variety of reaction interfaces. Seven different reaction zones can be identified according to the reactor model for ladle injection [1] (Figure 8.1).

The powder particles, on entering the gas stream, are accelerated by the high velocity of the carrier gas. At the lance tip, the gas expands as a result of a pressure drop and rise in temperature. If the velocity of the gas at the lance tip is sufficiently high then a gas jet cone forms, otherwise bubbling takes place. The transition from bubbling to jetting during powder injection has been studied in detail [2]. Under most industrial conditions in steelmaking jetting is likely to occur during powder injection.

On leaving the nozzle, the particles, because of their high momentum, travel almost in a straight line downwards until the gas jet breaks up into a swarm of rising gas bubbles as a result of the resistance offered by the metal and buoyancy forces. Some particles may penetrate into the liquid metal while the majority of them (over 60%), owing to frictional or surface tension forces, may become trapped inside the rising gas bubbles (or the rising gas plume). If, during their rise, the particles are wetted by the liquid metal, then they may break away from the bubble or be recirculated

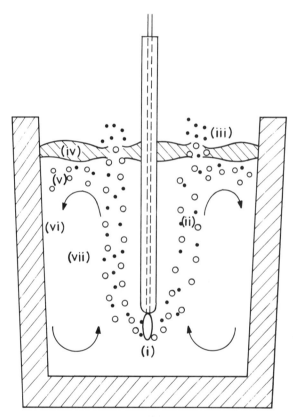

Figure 8.1 Reactor model for ladle injection. The different reaction zones are: (i) jet zone in front of lance outlet; (ii) plume zone containing bubbles and particles rising in metal; (iii) breakthrough zone where bubbles emerge; (iv) slag zone; (v) dispersion zone where gas–slag–metal emulsion can form; (vi) lining zone where metal may react with lining; and (vii) intermediate zone with lowest stirring intensity (adapted from Holappa [1]). (*Source*: Lehner, T. (1977) *Scaninject Proc. Int. Conf. on Injection Metallurgy, Luleå, Sweden, 9–10 June 1977*, Mefos, Luleå, paper 11.)

back into the metal after reaching the top surface. Some particles may coagulate, especially if melting of the particles occurs. A slag–metal–gas emulsion may form near the top surface, and bigger bubbles may break up into smaller ones. It is not easy to estimate the fraction of particles separated or recirculated back into the liquid metal from theoretical considerations alone. The thermodynamic and kinetic fundamentals of deoxidation and desulphurization have already been discussed in Chapters 2 and 3. The kinetic model of desulphurization discussed in this section has been developed for the injection of calcium carbide into carbon-containing molten iron. In principle, the same model can be applied to the desulphurization of liquid steel by powders like $CaO + CaF_2$, $CaO + Al_2O_3$, $CaO + CaF_2 + Al_2O_3$.

8.2.1   *Theoretical considerations*

The fundamental aspects of the treatment presented here are based on the work reported in Robertson *et al.* [3] and Deo and Grieveson [4].

The overall desulphurization rate during powder injection can be expressed as the sum of the rates of transitory ($Rate_t$) and permanent ($Rate_f$) reactions:

$$-V \frac{dc_m^b}{dt} = Rate_t + Rate_f \qquad (8.1)$$

where V is volume of metal (m^3) and the uniform concentration of sulphur in the bulk metal is given by c_m^b (mol/m^3).

By definition, the permanent reaction takes place due to the top slag alone, and the transitory reaction takes place while the injected powder particles rise in the plume. A small fraction of particles may also become entrained in the circulating stream of metal around the plume, or the particles may form an emulsion layer close to the slag–metal interface. These two mechanisms of desulphurization are not considered in this model because of the uncertainties involved in predicting the fraction of such particles and the associated area of reaction interfaces and the mass transfer coefficients (due to changing fluid flow fields).

Permanent reaction

Let the concentration in the slag and metal phases be represented as follows:

C_m^i = sulphur concentration at slag–metal interface on metal side (mol/m^3)

C_s^i = sulphur concentration at slag–metal interface on slag side (mol/m^3)

C_m^b, C_s^b = sulphur concentrations in bulk metal and bulk slag, respectively (mol/m^3)

If it is assumed that mass transfer in the boundary layer within the metal is rate controlling (i.e. $C_s^i = C_s^b$) and the bulk metal is thoroughly mixed, then

$$\text{Rate}_f = A_f k_f (C_m^b - C_m^i) \tag{8.2}$$

where A_f is the nominal area of slag–metal contact for permanent reaction (m²) and k_f is the mass transfer coefficient for permanent reaction (m/s). Let the partition coefficient L be defined as

$$L = C_s^i / C_m^i \tag{8.3}$$

In the case of metal phase control, $C_s^i = C_s^b$, and thus

$$L = C_s^b / C_m^i \tag{8.4}$$

The actual values of L can be determined from thermodynamic equilibrium calculations. On substituting Equation (8.4) into Equation (8.2)

$$\text{Rate}_f = A_f k_f (C_m^b - C_s^b / L) \tag{8.5}$$

Transitory reaction

During gas-borne powder injection through a vertical submerged lance, the gas jet (along with the powder particles) penetrates into the metal up to a certain depth before breaking down into a swarm of bubbles [5]. Various trajectories are possible for the powder particles. Some particles, type A, may sit inside the bubbles, partially or fully, while others, type B, may rise freely (or separately) inside the plume. Henceforth, the term 'freely rising particles (FRP)' will be used for particles of type B.

Let C_s^l be the average concentration of sulphur in FRP (the superscript l is used for FRP). If A_p denotes the total surface area of these particles (in m²) at any time t, then the overall equation for desulphurization (similar to Equation (8.5)) is

$$V_p^l \frac{dC_s^l}{dt} = A_p k_p (C_m^b - C_s^l / L) \tag{8.6}$$

where V_p^l is the total volume of FRP in the metal (m³) and k_p is the mass transfer coefficient in metal around the FRP (m/s). The maximum value of t is the residence time trp (in seconds) of the particles in the plume.

On rearrangement, Equation (8.6) yields

$$\frac{dC_s^l}{C_m^b - C_s^l / L} = \frac{A_p k_p}{V_p^l} dt \tag{8.7}$$

If, for the sake of simplicity, it is assumed that C_m^b remains constant during the time interval trp and the initial sulphur content of FRP is zero, then the integration of Equation (8.7) from $t = 0$ to $t = trp$ gives the final sulphur concentration in

FRP at the top of plume, $C_s^{l,f}$ (in mol/m³), just before mixing with bulk slag:

$$C_s^{l,f} = LC_m^b\left[1 - \exp\left(-\frac{A_p k_p\, trp}{V_p^l L}\right)\right]$$ (8.8)

Let A_g denote the total surface area of contact between the particles inside the bubbles (PIB) and the liquid metal and let C_s^g be the average concentration of sulphur in these particles. Then

$$V_p^g \frac{dC_s^g}{dt} = k_g A_g (C_m^b - C_s^g/L)$$ (8.9)

where V_p^g is the volume of particles inside the bubbles (PIB) (m³) and k_g is the mass transfer coefficient of sulphur in metal for PIB (m/s).

Integration of Equation (8.9) from $t = 0$ to $t = trb$ (residence time of bubbles in seconds) gives the average sulphur concentration in PIB at the top of the plume, $c_s^{g,f}$ (mol/m³):

$$C_s^{g,f} = LC_m^b\left[1 - \exp\left(-\frac{k_g A_g\, trb}{V_p^g L}\right)\right]$$ (8.10)

The rate of the transitory reaction is thus

$$\text{Rate}_f = (V_p^l/trp)C_s^{l,f} + (V_p^g/trb)C_s^{g,f}$$ (8.11)

The overall rate equation for desulphurization (Equation (8.1)) can now be written as

$$-V\frac{dC_m^b}{dt} = A_f k_f (C_m^b - C_s^b/L) + (V_p^l/trp)C_s^{l,f} + (V_p^g/trb)C_s^{g,f}$$ (8.12)

This equation is a general rate expression for transitory and permanent contact reactions occurring during powder injection in any slag/metal system. Since the value of L for slags of high sulphide capacity is large (i.e. $L > 2000$ for calcium carbide liquid slags [6]), the term C_s^b/L in Equation (8.12) can be neglected as a first approximation and Equation (8.12) can be combined with Equations (8.8), (8.10) and (8.11) to give

$$-V\frac{dC_m^b}{dt} = C_m^b\left\{A_f k_f + \frac{LV_p^l}{trp}\left[1 - \exp\left(-\frac{A_p k_p\, trp}{V_p^l L}\right)\right]\right.$$
$$\left. + \frac{LV_p^g}{trb}\left[1 - \exp\left(-\frac{A_g k_g\, trb}{V_p^g L}\right)\right]\right\}$$ (8.13)

The values of V_p^l, V_p^g, A_p and A_g can be computed as follows. Let f be the fraction of particles which reside inside the bubbles at the bubble–metal interface when the powder is injected. Then V_p^l and V_p^g are given by

$$V_p^l = \frac{Wtrp}{\rho_s}(1-f) \tag{8.14}$$

$$V_p^g = \frac{Wtrb}{\rho_s}f \tag{8.15}$$

where W is the rate of powder injection (kg/s) and ρ_{sl} is the density of slag (kg/m^3).

If d_p denotes the average particle diameter (in m), then for spherical particles the surface to volume ratio is given by $6/d_p$. Thus

$$A_p/V_p^l = 6/d_p$$

or

$$A_p = 6V_p^l/d_p \tag{8.16}$$

The area of PIB which contributes to desulphurization can be calculated as follows. If x is the diameter of the hemispherical cap bubble (m), then

$$A_g = m\,\frac{\pi x^2}{4}\,n_B \tag{8.17}$$

where n_B is the number of bubbles and m is a dimensionless factor called the 'effective area factor'.

If it is assumed that the particles in the bubble (PIB) are sitting at the base of the bubble, then the factor m is a measure of the fraction of the bubble base area which contributes to desulphurization. However, there is a limit on the maximum value of m because of a limit on the total bubble area for a given gas flow rate. The method of calculation of m is explained in Appendix E.

If d_B denotes the equivalent diameter of each hemispherical cap bubble (in m), then x and d_B are related by

$$\tfrac{2}{3}\pi(x/2)^3 = \tfrac{4}{3}\pi(d_B/2)^3$$

or

$$x = 2^{1/3}d_B \tag{8.18}$$

The number of bubbles can be estimated by assuming that heat transfer to the gas is fast, i.e. gas temperature equals bath temperature. If Q_{bath} denotes the gas flow rate (in m^3/s) at the temperature of the bath, Q_{stp} denotes the corresponding gas flow rate (in m^3/s) at stp and T_{bath} denotes the temperature of the bath (in K), then

$$Q_{bath} = Q_{stp}\,\frac{T_{bath}}{298} \tag{8.19}$$

where the carrier gas is assumed to behave ideally. The number of bubbles, n_B, is given by

$$n_B = \frac{Q_{bath} trb}{(\pi/6)d_B^3} \tag{8.20}$$

where the numerator denotes the volume of gas in the metal and $(\pi/6)d_B^3$ is the volume of each bubble. From Equations (8.17)–(8.20)

$$A_g = 2.38m \frac{T_{bath}}{298} trb \frac{Q_{stp}}{d_B} \tag{8.21}$$

Equation (8.13) can be rewritten as

$$-V \frac{dC_m^b}{dt} = C_m^b (a + b + c) \tag{8.22}$$

where

$$a = A_f k_f \tag{8.23}$$

$$b = \frac{LV_p^l}{trp} \left[1 - \exp\left(- \frac{A_p k_p trp}{V_p^l L} \right) \right] \tag{8.24}$$

$$c = \frac{LV_p^g}{trb} \left[1 - \exp\left(- \frac{A_g k_g trb}{V_p^g L} \right) \right] \tag{8.25}$$

The permanent reaction rate is proportional to the term a (Equation (8.23)), the reaction rate due to the FRP is proportional to the term b (Equation (8.24)) and the reaction rate due to PIB is proportional to the term c (Equation (8.25)). Thus, the various fractional contributions to the overall reaction can be expressed as follows:

$$\text{Permanent contribution,} \quad \text{PERC} = \frac{a}{a + b + c} \tag{8.26}$$

$$\text{Particle contribution,} \quad \text{PC} = \frac{b}{a + b + c} \tag{8.27}$$

$$\text{Bubble contribution,} \quad \text{BC} = \frac{c}{a + b + c} \tag{8.28}$$

Equation (8.22), on integration from $t = 0$ to $t = t_{in}$ (injection period in seconds) and on conversion of concentrations to wt %, gives for desulphurization ($C_m^b \rightarrow (\%S)$)

$$\ln\left(\frac{(\%S)_i}{(\%S)_f} \right) = (a + b + c) \frac{t_{in}}{V} \tag{8.29}$$

where the values of a, b and c are from Equations (8.14)–(8.25):

$$a = A_f k_f \qquad (8.30)$$

$$b = \frac{LW}{\rho_s}(1 - f)\left[1 - \exp\left(-\frac{6k_p trp}{d_p L}\right)\right] \qquad (8.31)$$

$$c = \frac{LW}{\rho_s} f\left[1 - \exp\left(-\frac{2.38}{d_B} m \frac{T_{bath}}{298} \frac{k_g trb Q_{stp}\rho_{sl}}{Wf L}\right)\right] \qquad (8.32)$$

These equations can be used to calculate the values of a, b and c for a given set of experimental data.

Example 8.1 **Desulphurization during the injection of calcium carbide in a 300 ton torpedo**

Let the typical values of different parameters (as at Hoogovens IJmuiden) be:
$k_f = 8.7 \times 10^{-4}$ m/s; $Q_{stp} = 2.2$ m³/min; $trb = trp = 0.75$ s; $d_p = 60$ μm; $d_B = 70$ mm; $\rho_s = 2000$ kg/m³; $W = 60$ kg/min; $k_p = k_g = 2.36 \times 10^{-3}$ m/s; $f = 0.7$; $A_f = 2.61 \times 2.61\pi$ m²; $t = 600$ s $T_{bath} = 1600$ K; $L = 2000$; $m = 1$.

On substituting the above values in Equations (8.26)–(8.32), we obtain

$$a = 0.018\,61 \quad b = 0.025\,41 \quad c = 0.016\,78$$

$$a + b + c = 0.0608$$

Hence if the sulphur content at the start of treatment is 0.02 wt %, then after 600 s the final sulphur content will be 0.0085 wt %. Also

$$\text{Permanent contribution} = \frac{a}{a + b + c} \times 100 = 30.6\%$$

$$\text{Particle contribution} \quad = \frac{b}{a + b + c} \times 100 = 41.8\%$$

$$\text{Bubble contribution} \quad = \frac{c}{a + b + c} \times 100 = 27.6\%$$

8.3 Alloy dissolution

Metals and alloys can be added to liquid steel at various stages of steelmaking (i.e. inside the furnace, during furnace tapping, during vacuum treatment and/or argon stirring) so as to obtain the desired composition of the melt. The timing of the addition depends upon the nature of the addition (i.e. its melting point, oxidation characteristics, volatility) and the process route followed. Nickel can be added in the electric arc furnace at any time as NiO, because the latter is easily reduced. Since the reduction of Cr_2O_3 from slag to metal is difficult, ferro chrome additions are made when top slag is reducing (low FeO content). Molybdenum has a high vapour pressure and ferro molybdenum is added in the electric furnace just before tapping to minimize losses due to vaporization. In the oxygen steelmaking process route some alloying additions (like Fe–Si, Fe–Mn) are made during furnace tapping and the rest are added during subsequent stages of secondary steelmaking, i.e. during argon stirring, vacuum treatment, etc.

There are two principal classes of ferro alloys. The melting point of class I ferro alloys – like Fe–Si, Fe–Mn, Si–Mn, Fe–Ca – is lower than the temperature of liquid steel. As soon as the ferro alloy is added to liquid steel a solidified shell of steel may form around the ferro alloy particle due to the local freezing effect. With time, the shell melts and the ferro alloy particle inside the shell is also heated up to its melting point. The complete dissolution time of class I ferro alloys depends upon the heat transfer coefficient in the molten steel bath as well as on the size of the ferro alloy added.

The alloys of molybdenum, vanadium, niobium and tungsten which have a melting point higher than that of liquid steel are designated as class II ferro alloys. The dissolution kinetics of class II alloys is slower than that of class I alloys; the rate of dissolution is controlled by mass transfer in the liquid steel, even in an agitated bath. It is therefore important to have good control of the size range (3–10 mm) of class II ferro alloy additions so as to obtain good mixing, fast dissolution and high recovery. Powdered compacts of ferro alloys like Fe–V, Fe–W and Fe–Mo dissolve more rapidly compared with equivalent-sized solid pieces. Autoexothermic alloys (which generate heat on melting) can also be used for faster melting, better recovery and uniform distribution.

The thermophysical properties of class I and class II ferro alloys are summarized in Tables 8.1 and 8.2 respectively [7].

Several methods of alloy addition are practised in industry, such as simple throwing of filled bags, throwing with a shovel, addition through mechanized chutes, wire feeding, powder injection, bullet shooting, dipping stationary alloy ingots in an agitated bath and the so-called Composition Adjustment by Sealed argon bubbling (CAS) process. In the CAS process, a refractory-lined snorkel is partially immersed in the liquid steel in the ladle such that it envelops the rising gas plume (created by the injection of argon gas through the porous plug at the bottom). Alloy additions are made on the liquid metal surface within the area covered by the snorkel. Inert gas coming out through the plume eye within the snorkel has a low partial pressure of oxygen and thus prevents oxidation of the added alloy by the ambient atmosphere. Melting and distribution rates are high as a result of the agitation induced by the rising gas plume.

The governing equations for fluid flow during axisymmetric gas injection as well as their method of solution have been briefly described in Chapter 1. If the velocity distribution in the bath is known, then for class II ferro alloys we can estimate the mass transfer coefficient in the metal phase by using one of the several correlations

Table 8.1   Physical and thermal properties relevant to class I ferro alloys.

Material	Density (kg/m³)	Heat capacity (J/kg K)	Thermal conductivity (W/m K)	Latent heat (J/kg)	$T_{solidus}$ (K)	$T_{liquidus}$ (K)
Ferro manganese Mn = 79.5% C = 6.4% Si = 0.27% Fe: balance	7200.0	700.0	7.53	534 654.0	1344.0	1539.0
Silico manganese Mn = 65.96% Si = 17.07% C = 1.96% Fe: balance	5600.0	628.0	6.28	578 783.0	1361.0	1489.0
Ferro silicon 50% Si = 49.03% Al = 1.20% max. Fe: balance	4460.0	586.0	9.62	908 200.0	1483.0	1500.0
Ferro chrome Cr = 50–58% C = 0.25% max. Si = 1.5% max. Mn = 0.50% max. Al = 1.50% max. Fe: balance	6860.0	670.0	6.50	324 518.0	1677.0	1755.0

Table 8.2   Physical and thermal properties relevant to class II ferro alloys.

Material, A	Density (kg/m³) (1873 K)	Heat capacity (J/kg K)	Thermal conductivity (W/m K)	Diffusivity ($D_{A/Fe} \times 10^9 \, m^2/s$)
Molybdenum	10 000	310	100	3.2
Vanadium	5700	400	50	4.1
Niobium	8600	290	64	4.6
Tungsten	19 300	140	115	5.9

[8] given for specific cases in Table 8.3. If the initial concentration of alloy in the bulk is negligible then the rate of dissolution can be expressed as

$$\frac{dR_s}{dt} = k \, \frac{C_s^i}{\rho_{alloy}} \tag{8.33}$$

where R_s is the radius of the added alloy particle, k is the mass transfer coefficient, C_s^i is the interface concentration of alloy (namely, of molybdenum in Fe–Mo) and ρ_{alloy} is the density of alloy. Numerically predicted dissolution rates of three different

Table 8.3 Typical correlations for solid–liquid mass transfer (adapted from Lange [8]).

Sl. No.	Correlation	Ref.	Remarks
1	$Sh = 0.05 Re^{0.74} Sc^{1/3}$	(a)	Dissolution of carbon rods at plume–slag contacting edge
2	$Sh = 0.016 Re^{0.75} Sc^{1/3}$	(b)	For dissolution of rotating solid steel
3	$Sh = 0.112 Re^{0.67} Sc^{0.356}$	(c)	
4	$Sh = 0.104 Re^{0.70} Sc^{1/3}$	(d)	Dissolution of Cr_2O_3 rods in Fe–C or Fe–10 Cr–C alloys
5	$Sh = 2 + 0.55 Re_p^{0.79} Sc^{1/3}$	(e)	For bubble columns
6	$Sh = 2 + 0.45 Re_p^{0.76} Sc^{1/3}$	(f)	For bubble columns and aerated vessels

Sources:
(a) T. Lehner, G. Carlsson and H. Tse-Chiang (1980) *Scaninject II*, Mefos: Luleå 22/1–22/34.
(b) M. Kosaka and S. Minowa (1967) *Tetsu-to-Hagané*, **53**, 983–97.
(c) Y. U. Kim and R. D. Pehlke (1974) *Metall. Trans.*, **5B**, 2527–32.
(d) K. Suzuki, K. Mori and T. Ito (1979) *Tetsu-to-Hagané*, **65**, 1131–9.
(e) P. Songer and W. D. Deckwer (1981) *Chem. Eng. J.*, **22**, 179–86, $Re_p = (d_p^{4/3} E^{1/3})/v_1$ where d_p is particle diameter (m), E is power dissipated per unit mass of liquid (m^2/s^3) (which is average power input to the fluid per unit mass in W/kg), v_1 is the kinematic viscosity of liquid (m^2/s).
(f) G. M. Marrone and D. J. Kirwan (1986) *Am. Inst. Chem. Eng. J.*, **23**, 523–5.

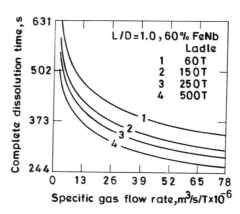

Figure 8.2   Numerically predicted complete dissolution time of a typical class II ferro alloy in ladles of various sizes (height to diameter ratio unity) as a function of specific gas flow rate. (*Source*: Mazumdar, D. (1992) *AIME Proc. Int. Conf. on Process Modelling*, Toronto).

class II alloys in a 60 ton argon-stirred ladle at various gas flow rates [9] are shown in Figure 8.2. Typical complete dissolution times are of the order of 10–15 minutes. It is predicted that in larger-capacity ladles, the dissolution times are smaller for the same rate of specific gas consumption and length to diameter ratio of ladle.

8.4 Inclusion modification by calcium treatment of liquid steel

The injection of calcium and its alloys into liquid steel has assumed great importance for sulphur, oxygen and inclusion control so as to produce quality steels with improved physical and mechanical properties.

Calcium metal has a melting point of 850 °C and a boiling point of 1490 °C. At steelmaking temperatures, calcium melts, vaporizes and then dissolves into liquid steel:

$$\langle Ca \rangle = \{Ca\}$$

$$\{Ca\} = (Ca)_g$$

$$(Ca)_g = [Ca]$$

Dissolved calcium reacts with dissolved oxygen (or oxygen present as Al_2O_3) and dissolved sulphur (or sulphides such as MnS):

$$[Ca] + [O] = (CaO)$$

$$[Ca] + [S] = (CaS)$$

$$[Ca] + (x + \tfrac{1}{3})(Al_2O_3) = (CaO \cdot x\,Al_2O_3) + \tfrac{2}{3}[Al]$$

A range of calcium aluminates may form in the following sequence depending upon the concentrations of [Ca], (Al_2O_3), [O] and [Al]:

$$Al_2O_3 \rightarrow CaO \cdot 6Al_2O_3 \rightarrow CaO \cdot 2Al_2O_3 \rightarrow CaO \cdot Al_2O_3 \rightarrow 12CaO \cdot 7Al_2O_3$$

At 1600 °C, the well-established reaction equilibria are as follows [10]:

$$\frac{[Al]^2[O]^3}{a_{Al_2O_3}} = 1.2 \times 10^{-13}$$

$$\frac{a_{CaS}[O]}{a_{CaO}[S]} = 0.053$$

In lime-saturated slags $a_{CaS} = 1$, $a_{Al_2O_3} = 0.005$; in 50:50 CaO:Al_2O_3, $a_{CaO} = 0.34$ and $a_{Al_2O_3} = 0.064$; in aluminate-saturated slags $a_{CaO} = 0.11$ and $a_{Al_2O_3} = 0.30$.

Solubility data of calcium in metal are doubtful. There is a large uncertainty reported in the solubility product: $[Ca] \times [O] = 10^{-6}$ to 10^{-12}. The vapour pressure of calcium at 1600 °C may vary between 1.6 and 3.7 atm. Taking the solubility product $[Ca] \times [O] = 10^{-7}$, as reported in Gatellier *et al.* [11], the thermodynamic equilibrium relations between Ca–S–O–Al are presented in Figure 8.3. The recovery of calcium in metal during injection treatment is less than 10% because most of the calcium escapes in vapour form. In order to increase the dissolution time available, calcium is alloyed with silicon and Ca–Si alloy powder is injected. The uncertainties in thermodynamic data and the fluctuation in percentage recovery of calcium during calcium treatment have made the theoretical prediction of the kinetics of reactions involving dissolved calcium extremely difficult. Practical guidelines are used and the practice to be followed at a plant is mostly decided by way of trials.

The activity of Al_2O_3 in calcium aluminate is low. Therefore, as shown in Figure 8.4, in aluminium-killed steels the dissolved oxygen activity is less by 3–6 ppm when the steel is either in contact with calcium aluminate slag or subjected to calcium treatment (i.e. calcium aluminate precipitates within liquid steel) [10]. The primary objective of calcium treatment is to form liquid calcium aluminate. The liquid particles, in contrast to solid Al_2O_3, can grow in size and float out of the liquid steel easily, leaving behind a cleaner steel with low oxide contents. Besides Al_2O_3, it is important to control dissolved sulphur in steel because part of the calcium is used up in forming calcium sulphide. It is found that in liquid steels containing less than 40 ppm sulphur, there is almost complete conversion of solid Al_2O_3 to molten calcium aluminate [8]. In special grades sulphur could be as low as 10 ppm. The influence of ladle refractory

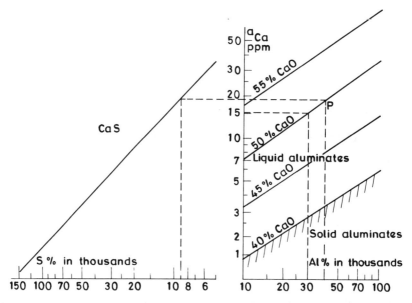

Figure 8.3 Equilibrium diagram for Fe–Al–Ca–O–S system at 1600 °C. (*Source:* Gatellier, C., Gaye, H. and Nadif, M. (1988) *Proc. Int. Calcium Treatment Symp. June 30, 1988, Glasgow,* pp. 31–37.)

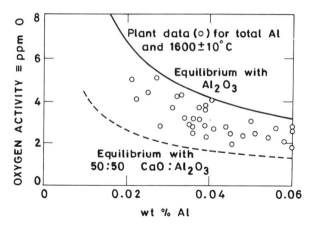

Figure 8.4 Equilibrium data and plant data for deoxidation with aluminium and 50:50 $CaO:Al_2O_3$ molten slag during furnace tapping. (*Source:* Turkdogan, E. T. (1988) *Proc. Int. Calcium Treatment Symp., June 30, 1988, Glasgow*, pp. 3–13.)

lining on desulphurization efficiency during Ca–Si injection is shown in Figure 8.5 [1]; dolomite lining is preferred for achieving very low sulphur contents.

Yet another important aspect of calcium treatment, in addition to the formation of liquid calcium aluminate, is the modifying effect of calcium on inclusion morphology, schematically shown in Figure 8.6 [1]. A complete modification of type II manganese sulphide (elongated in shape) inclusions takes place when the sulphur content is less than 0.07%. In practice 0.5–0.7 kg of Ca–Si per ton (of steel) is necessary for inclusion shape control [8]. All particles may not float out. Even if the

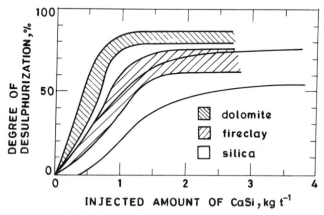

Figure 8.5 Influence of ladle refractory material on desulphurization efficiency (adapted from Holappa [1]). (*Source:* Förster, E., Klapdar, W., Richter, H., Rommerswinkel, H.-W., Spetler, E. and Wendorff, J. (1974) *Stahl und Eisen*, **94**, 474–85.)

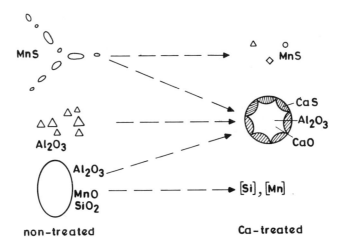

Figure 8.6   Schematic representation showing modification of inclusions with calcium treatment. (*Source*: Tähtinen, K. Väinölä, R. and Sandholm, R. (1980) *Scaninject II, Proc. 2nd Int. Conf. on Injection Metallurgy, Luleå, Sweden, 12–13 June 1980*, Mefos, Luleå, paper 24.)

small globular particles (covered with a layer of CaS, Figure 8.6) are unable to float out they are found to be less detrimental to the physical and mechanical properties of steel. Detailed studies of the effect of inclusions on the physical and mechanical properties of steel (not included in this book) are available [1, 8].

Some practical consequences of calcium treatment are as follows. Almost all the silicon from Ca–Si is dissolved in liquid steel (100% recovery of silicon). Phosphorus in the slag may revert from the slag to the metal and therefore it is important to minimize slag carry over from the furnace to the ladle. Oxidation loss of dissolved aluminium and titanium is much less during Ca–Si injection than during the powdered injection of lime-based fluxes (i.e. injection of $CaO + CaF_2$); hydrogen pickup is also comparatively less during Ca–Si injection but, unless special care is taken, nitrogen pick up can increase by 10–30 ppm. The castability of calcium-treated steel is better; nozzle clogging (caused by the deposition of Al_2O_3 clusters on the nozzle surface) is reduced but care should be taken so that liquid aluminates form; there may not be enough calcium to modify Al_2O_3 particles below 0.4 kg of calcium per ton (of steel) (or below a Ca/Al ratio of 0.14); above 1 kg of calcium added per ton (of steel) there is a risk of forming solid calcium aluminates [8].

Two methods are used for injecting calcium into the metal. In gas-borne powdered injection with the help of a submerged lance, Ca–Si particles (0.6–1 mm diameter) are injected at a carrier gas pressure of 5–15 atm. (A kinetic model for powder injection is developed in section 8.2). An alternative to powder injection is wire feeding in which the powdered reagent is encased in a metal sheath (commercially available as a wire product) and continuously fed deep into the metal. The choice between powder injection and wire feeding mainly depends upon economic and practical considerations. The recovery of calcium in wire feeding may be three to six times higher than that obtained during Ca–Si powder injection. The initial capital investment for a powder injection installation is much more than that for a wire feeding installation but cored wires are approximately three times more expensive than ordinary powder per kilogram of calcium.

References

1. Holappa, L. E. K. (1982) 'Ladle injection metallurgy', *Int. Mater. Rev.*, **27**, 53–76.
2. Farias, L. R. and Irons, G. A. (1985) 'A unified approach to bubbling–jetting phenomena in powder injection into iron and steel', *Metall. Trans.*, **16B**, 211–25.
3. Robertson, D. G. C., Ohguchi, S., Deo, B. and Willis, A. (1983) 'Theoretical and laboratory studies on injection phenomenon', *Scaninject III*, Part 1, Mefos: Luleå, paper 8:1–47.
4. Deo, B. and Grieveson, P. (1988) 'Desulphurization of molten pig iron containing aluminium by powder injection', *Steel Res.*, **59**, 237–46.
5. Irons, G. A. (1989) 'Role of injection in powder desulphurization process', *Ironmaking and Steelmaking*, **16**, 28–36.
6. Schürmann, E. and Delhey, H. M. (1990) 'Thermodynamics of desulphurization reactions during treatment of hot metal with calcium and calcium compounds', *Steel Res*, **61**, 64–71.
7. Mazumdar, D. (1990) 'Alloying of steel', *Short-term Course on Secondary Steelmaking, December 1990, IIT, Kanpur*, Chapter 6, pp. 69–80.
8. Lange, K. W. (1988) 'Thermodynamic and kinetic aspects of secondary steelmaking processes', *Int. Mater. Rev.*, **33**, 53–89.
9. Mazumdar, D. (1992) 'Mathematical modelling of ferro alloy dissolution in argon-stirred ladles', Private communication with B.D., IIT, Kanpur.
10. Turkdogan, E. T. (1988) 'Metallurgical consequences of calcium retention in liquid and solid steel', *Proc. Int. Calcium Treatment Symp., June 30, 1988, Glasgow*, pp. 3–13.
11. Gatellier, C., Gaye, H. and Nadif, M. (1988) 'Provision of inclusions composition in calcium treated steel', *Proc. Int. Calcium Treatment Symp., June 30, 1988, Glasgow*, pp. 31–7.

Absolute and collision theories of reaction rate

A.1 Reversible and irreversible reactions

Consider the reaction $A + B = C + D$:

$$\text{Rate of forward reaction} = k_f C_A C_B \qquad (A.1)$$

$$\text{Rate of backward reaction} = k_b C_C C_D \qquad (A.2)$$

$$\text{Net rate per unit area } (r) = k_f C_A C_B - k_b C_C C_D \qquad (A.3)$$

At equilibrium $r = 0$, and if the corresponding equilibrium concentrations are C_C^e, C_D^e, C_A^e, C_B^e, then

$$k_f C_A^e C_B^e = k_b C_C^e C_D^e \qquad (A.4)$$

or

$$\frac{k_f}{k_b} = \frac{C_C^e C_D^e}{C_A^e C_B^e} = K_e \qquad (A.5)$$

where K_e is the equilibrium constant.

Combining Equations (A.3) and (A.5):

$$r = k_f\left(C_A C_B - \frac{1}{K_e} C_C C_D\right) \qquad (A.6)$$

Equation (A.6) represents the rate expression for a reversible reaction, i.e. a reaction taking place under conditions not far from equilibrium. For an irreversible reaction which takes place far away from equilibrium, the backward reaction can be ignored, i.e.

$$r = k_f C_A C_B \qquad (A.7)$$

A.2 Order of reaction

The order of reaction is defined as the total power to which concentration is raised in the experimental rate law; for example, in Equation (A.7) the powers of C_A and C_B are unity and hence the order of reaction is $1 + 1 = 2$. If the reaction rate is constant and hence independent of concentration, then it is called a zero-order reaction. Because most of the reactions in steelmaking take place at high temperature, they are controlled by mass transfer and not by chemical reaction; mass transfer control reactions, in terms of the rate equation, are first-order reactions.

A.3 The Arrhenius equation

The variation of equilibrium constant with temperature is given by

$$\frac{d(\ln K_e)}{dT} = \frac{\Delta H_T^{o, react}}{RT^2} \tag{A.8}$$

where $\Delta H_T^{o, react}$ is the enthalpy of the reaction. Substituting Equation (A.5) into (A.8)

$$\frac{d(\ln k_f)}{dT} - \frac{d(\ln k_b)}{dT} = \frac{\Delta H_T^{o, react}}{RT^2} \tag{A.9}$$

which may be split into the following equations:

$$\frac{d(\ln k_f)}{dT} = \frac{E_1}{RT^2} \tag{A.10}$$

and

$$\frac{d(\ln k_b)}{dT} = \frac{E_2}{RT^2} \tag{A.11}$$

where $E_1 - E_2 = \Delta H_T^{o, react}$.

If the energy terms are assumed to be independent of temperature, integration of Equation (A.10) or (A.11) gives the form

$$k = A \exp(-E/RT) \tag{A.12}$$

This is the Arrhenius equation. A is a pre-exponential term (frequency factor) and E is the activation energy. The value of E can be obtained from the slope of the plot of $\ln k$ versus $1/T$.

A.4   Collision theory of reaction rate

The collision rate of molecules in a gas can be found from the kinetic theory of gases. For the bimolecular collisions of like molecules of A, we have

$$Z_{AA} = \text{number of collisions of A with A (per second, per m}^3\text{)}$$

$$= d_A^2 n_A^2 \left[\frac{4\pi k_B T}{M_A} \right]^{1/2} = d_A^2 \frac{N^2}{10^6} \left[\frac{4\pi k_B T}{M_A} \right]^{1/2} C_A^2 \tag{A.13}$$

where

$$d_A = \text{diameter of molecule A, m}$$

$$M_A = \text{mass of molecule A, kg}$$

$$N = \text{Avogadro's number, } 6.023 \times 10^{23}$$

$$C_A = \text{concentration of A, kmol/m}^3$$

$$n_A = N C_A / 10^3 \text{ number of molecules of A (per m}^3\text{)}$$

$$k_B = R/N = 5.76 \times 10^{-16} \text{ (Boltzmann's constant, J/K)}$$

For the bimolecular collisions of unlike molecules in a mixture of A and B, the number of collisions of A and B, Z_{AB}, is given by

$$Z_{AB} = \left(\frac{d_A + d_B}{2} \right)^2 n_A n_B \left[8\pi k_B T \left(\frac{1}{M_A} + \frac{1}{M_B} \right) \right]^{1/2}$$

$$= \left(\frac{d_A + d_B}{2} \right)^2 \frac{N^2}{10^6} \left[8\pi k_B T \left(\frac{1}{M_A} + \frac{1}{M_B} \right) \right]^{1/2} C_A C_B \tag{A.14}$$

where M_B is the mass of molecule B, C_B is the concentration of B, d_B is the diameter of molecule B and n_B is the number of molecules of B per m^3.

If every collision between the reactant molecules results in the transformation of the reactants into the product, these expressions give the rate of a bimolecular reaction. The actual rate is usually much lower than the predicted rate indicating that only a small fraction of all collisions between reactant molecules result in reaction. This suggests that only the more energetic and violent collisions between molecules lead to reaction, or more specifically, only those collisions that involve energies in excess of a given minimum energy E lead to reaction. From the Maxwell distribution law of molecular energies, the fraction of all bimolecular collisions that involve energies in excess of this minimum energy is given by $\exp(-E/RT)$ when $E \gg RT$.

Thus the rate of reaction r_{AB} is

$$-r_{AB} = k C_A C_B$$

$$= \left(\begin{array}{c} \text{collision rate,} \\ \text{kmol/m}^3\text{ s} \end{array} \right) \left(\begin{array}{c} \text{fraction of collisions involving} \\ \text{energies in excess of } E \end{array} \right)$$

$$= Z_{AB} \frac{1000}{N} \exp(-E/RT)$$

$$= \left(\frac{d_A + d_B}{2}\right)^2 \frac{N}{10^3} \left[8\pi k_B T \left(\frac{1}{M_A} + \frac{1}{M_B}\right)\right]^{1/2} C_A C_B \exp(-E/RT) \qquad \text{(A.15)}$$

Thus for the reaction rate constant k

$$k \propto T^{1/2} \exp(-E/RT) \qquad \text{(A.16)}$$

A.5 The absolute reaction rate theory (or transition state theory)

When a reactant goes to product(s), it has to cross an energy barrier. There may be an infinite number of paths, but the path which requires the minimum energy will be the most favourable. This does not mean that there is only one path. The probability of a particular path will depend on energetics and is given by

$$\exp(-E/RT) \qquad \text{(A.17)}$$

The larger the value of E, the smaller will be the probability of that path. In absolute reaction rate theory we neglect all other paths and consider only that path which has the least energy barrier.

We make the following assumptions:

1. Reactants combine to form unstable intermediates called activated complexes.
2. Equilibrium exists between reactants and the activated complexes at all time.
3. Activated complexes decompose spontaneously into products. The rate of decomposition of the complexes is the same for all reactions and is given by $k_B T/h$ where k_B is Boltzmann's constant and h is Planck's constant.

Thus for the elementary reversible reaction

$$A + B \underset{k_2}{\overset{k_1}{\rightleftharpoons}} AB, \quad \Delta H_r \qquad \text{(A.18)}$$

we have the following conceptual scheme:

$$A + B \underset{k_4}{\overset{k_3}{\rightleftharpoons}} AB^* \overset{k_5}{\to} AB \qquad \text{(A.19)}$$

with $K_e^* = k_3/k_4 = AB^*/A.B$ and $k_5 = k_B T/h$. The observed rate of reaction is

$$r_{AB} = \left(\begin{array}{c}\text{concentration of} \\ \text{activated complex}\end{array}\right) \left(\begin{array}{c}\text{rate of decomposition} \\ \text{of activated complex}\end{array}\right)$$

$$= (k_B T/h)[AB^*]$$

$$= (k_B T/h)K_e^*[A][B] \tag{A.20}$$

By expressing the equilibrium constant of the activated complex in terms of the standard free energy

$$\Delta G^* = \Delta H^* - T\Delta S^* = -RT \ln K_e^*$$

or

$$K_e^* = \exp(-\Delta G^*/RT) = \exp(-\Delta H^*/RT + \Delta S^*/R) \tag{A.21}$$

The rate becomes

$$r_{AB} = \frac{k_B T}{h} \exp(\Delta S^*/R)\exp(-\Delta H^*/RT)[A][B] \tag{A.22}$$

Of the three terms that make up the rate constant in the above equation $\exp(\Delta S^*/R)$ is much less sensitive to temperature than the other two terms and so we can take it as a constant. Hence for the forward and reverse reactions of Equation (A.18) we find approximately that

$$k_1 \propto T \exp(-\Delta H_1^*/RT) \tag{A.23}$$

$$k_2 \propto T \exp(-\Delta H_2^*/RT) \tag{A.24}$$

where

$$\Delta H_1^* - \Delta H_2^* = \Delta H_T^{o,\,react}$$

Using analogous arguments from thermodynamics it can be shown for the Arrhenius activation energy E that

$$E = \Delta H^* + RT \tag{A.25}$$

(for liquids and solids) and

$$E = \Delta H^* - (\text{molecularity} - 1)RT$$

(for gases). Since $E \gg RT$

$$E = \Delta H^*$$

$$k \propto T \exp(-E/RT) \tag{A.26}$$

A.6   Comparison between the collision and absolute reaction rate theories

Consider A and B colliding and forming an unstable intermediate which then decomposes into the product or

$$A + B \rightarrow AB^* \rightarrow AB \tag{A.27}$$

Collision theory views the rate to be governed by the number of energetic collisions between reactants. It is simply assumed that the intermediate breaks down rapidly enough into products such that it does not influence the rate of the overall process.

The absolute reaction rate theory assumes that the reaction rate is governed by the rate of decomposition of the intermediate. The rate of formation of the intermediate is assumed to be so rapid that it is present in equilibrium concentrations at all times. The manner in which the intermediate forms is not an important concern. Note the difference between Equation (A.26) and Equation (A.16) regarding the effect of temperature T on k.

B Summary of models for nitrogen adsorption and desorption

The summary of models presented below is essentially based on extensive reviews published in the literature; see [11] and [12] in Chapter 3. According to the latter reference, no convincing mechanism has yet been established for nitrogen adsorption and desorption processes because the information available in the literature is either insufficient or conflicting. The models briefly discussed below merely demonstrate the variety of approaches possible in nitrogen–metal reactions.

B.1 Surface blockage model

If θ_s is the surface sites blocked by a surfactive element, the apparent rate constant (k_{ap}) can be written as

$$k_{ap} = k_c(1 - \theta_s) \tag{B.1}$$

where k_c is the rate constant for metal free of the surfactive element. As $\theta_s \to 1$:

$$1 - \theta_s = K_s/a_s \tag{B.2}$$

where K_s is the adsorption coefficient (wt %) and a_s is the activity of solute, approximated by wt %.

From Equations (B.1) and (B.2)

$$k_{ap} = \frac{k_c K_s}{a_s} \tag{B.3}$$

i.e. with an increase in a_s the value of k_{ap} would decrease. Since the rate of forward reaction is given by

$$\frac{dC_N^b}{dt} = \left(\frac{A}{V}\right)(1 - \theta_s)k_c p_{N_2}$$

On substituting θ_s we obtain

$$\frac{dC_N^b}{dt} = \left(\frac{A}{V}\right)\left(\frac{k_c K_s}{a_s}\right)p_{N_2} = K_{ap}\frac{A}{V}\,p_{N_2} \tag{B.4}$$

where C_N^b is the bulk concentration of nitrogen, according to which, at high surface coverage,

$$\frac{dC_N^b}{dt} \propto p_{N_2} \tag{B.5}$$

B.2 Dissociative adsorption control model

The rate-controlling step is assumed to be the dissociative adsorption

$$(N_2)_g \Leftrightarrow 2(N) \tag{B.6}$$

If the rates of the forward and backward (reverse) reaction are respectively k_f and k_r,

$$\frac{dC_N^b}{dt} = \frac{A}{V}\left[k_f p_{N_2} - k_r (C_N^b)^2\right] \tag{B.7}$$

Since, at equilibrium,

$$\frac{k_f}{k_r} = \frac{(C_N^{eq})^2}{p_{N_2}} \tag{B.8}$$

we get

$$\frac{dC_N^b}{dt} = \frac{A}{V}\,k_r\left[(C_N^{eq})^2 - (C_N^b)^2\right] \tag{B.9}$$

which implies a second-order reaction.

B.3 Sublayer model

It is assumed that at high concentrations the surfactive elements form a sublayer beneath the surface through which nitrogen diffusion becomes rate controlling. If C_N^s is the surface concentration of adsorbed nitrogen and C_N^i is the concentration at the sublayer(SL)/boundary layer (BL) junction, then the rate equation is given by

$$\left(\frac{dC_N^b}{dt}\right)_{SL} = k_{SL}\left(\frac{A}{V}\right)(C_N^s - C_n^i) \qquad C^* = \frac{k_{BL}}{k_{SL}}\,C_N^s \tag{B.10}$$

The rate of transfer in the liquid metal boundary layer below the sublayer is

$$\left(\frac{dC_N^b}{dt}\right)_{BL} = k_{BL}\left(\frac{A}{V}\right)(C_N^i - C_N^b) \tag{B.11}$$

where C^b is the bulk concentration of nitrogen.

Equating (B.10) and (B.11) to eliminate C_N^i:

$$\frac{dC_N^b}{dt} = k'\left(\frac{A}{V}\right)(C^* - C_N^b) \tag{B.12}$$

where

$$\frac{1}{k'} \text{ (total resistance)} = \frac{1}{k_{BL}} + \frac{1}{k_{SL}} \text{ (sum of the phase resistances)} \tag{B.13}$$

Since $C^* \propto \sqrt{(p_{N_2})}$, it shows that $dC_N^b/dt \propto \sqrt{(p_{N_2})}$ which is similar to the oxygen barrier model proposed by Pehlke and Elliott ([13] in Chapter 3).

B.4   Electrostatic double-layer model

It is assumed that in the presence of surfactive sulphur and oxygen the surface consists of an electrostatic double layer (DL) composed of S^{2-}, O^{2-} and Fe^{2+} ions through which nitrogen diffusion is rate controlling.

The rate equation is given by

$$\frac{dC_N^b}{dt} = \frac{A}{V} k_{DL}(\text{wt \% S})^{-2/3}(C_N^{eq} - C_N^b) \tag{B.14}$$

where C_N^{eq} is equilibrium nitrogen content. The term k_{DL} includes p_{N_2} terms.

B.5   Liquid-phase diffusion control model

Following the procedure given in Example 3.5, we can write

$$\ln\left(\frac{C_N^t - C_N^{eq}}{C_N^o - C_N^{eq}}\right) = k_m \frac{A}{V} t \frac{C_N^o}{C_N^o - C_N^{eq}} \tag{B.15}$$

where C_N^o is the initial nitrogen concentration and C_N^t is the nitrogen concentration at time t. Liquid-phase diffusion control has been proposed to hold good for the absorption of nitrogen in pure metal at low surfactive concentrations.

B.6 Activation model

It is assumed that an activated nitrogen molecule is formed at the surface due to collision with another molecule and the overall process of nitrogen adsorption proceeds according to

$$\tfrac{1}{2}N_2 \underset{k_2}{\overset{k_1}{\rightleftarrows}} \tfrac{1}{2}N_2^* \tag{B.16}$$

and

$$\tfrac{1}{2}N_2^* \overset{k_3}{\to} [N] \tag{B.17}$$

The rates of the reactions are given respectively by r_1 and r_2:

$$r_1 = k_1 p_{N_2} - k_2 p_{N_2^*}^{1/2} p_{N_2}^{1/2} \tag{B.18}$$

$$r_2 = k_3 p_{N_2^*}^{1/2} \tag{B.19}$$

At steady state

$$r_1 = r_2 = r = k_1 p_{N_2} \left(\frac{k_2}{k_3} p_{N_2}^{1/2} + 1 \right)^{-1} \tag{B.20}$$

For a slow dissolution process (i.e. at high surfactant concentration)

$$\left(\frac{k_2}{k_3} \right) p_{N_2}^{1/2} \ll 1 \tag{B.21}$$

and

$$r = k_1 p_{N_2} \tag{B.22}$$

For a fast dissolution process

$$\left(\frac{k_2}{k_3} \right) p_{N_2}^{1/2} \gg 1 \tag{B.23}$$

$$r = \left(\frac{k_1 k_3}{k_2} \right) p_{N_2}^{1/2} \tag{B.24}$$

The activation model thus tries to explain the dependence of rate as being directly proportional to nitrogen pressure (Equation (B.22)) at high surfactant concentration and to the square root of nitrogen pressure at low surfactant concentration (Equation (B.24)).

B.7   Mixed control models

Several types of mixed control models have been proposed depending upon the choice of the following assumption(s):

1. The overall process rate is controlled by surface reaction and mass transfer in the melt.

2. The overall process rate (R) is the sum of the rate of reaction through bare surface (R_{bs}) and the rate of reaction through covered sites (R_s):

$$R = (1 - \theta)R_{bs} + R_s \tag{B.25}$$

where θ represents the fraction of covered surface (by oxygen or sulphur) and can be evaluated by using a modified version of Langmuir's isotherm. The final expression for the rate is of the type

$$R = a + b(c + C_s)^{-1} \tag{B.26}$$

where a, b and c are constants and C_s is the surfactant concentration. Rao and Lee ([12] in Chapter 3) showed that the above equation could be empirically fitted to a large number of experimental data.

3. The overall process rate is governed by mixed gas phase mass transfer plus chemical reaction plus mass transfer in the metal phase.

Models 1 and 3 are termed sequential reaction models and model 2 is termed a parallel reaction model. According to Battle and Pehlke ([11] in Chapter 3) model 1 is promising because the rate constants involved can be determined unambiguously from the experimentally concentration–time data. The sequential rate equation based on mass transfer in metal (k_m) and chemical reaction at the interface (k_{CR}) is

$$\frac{dC_N^b}{dt} = k_{CR}\frac{A}{V}\left[(C_N^{eq})^2 - (C_N^i)^2\right] = k_m\frac{A}{V}(C_N^i - C_N^b) \tag{B.27}$$

where

C_N^b = bulk concentration

C_N^i = interfacial concentration

C_N^{eq} = nitrogen concentration in melt in equilibrium with nitrogen gas

C_N^i can be eliminated by solving Equation (B.27) to give

$$\frac{dC_N^b}{dt} = k_x\frac{A}{V}(C_N^{eq} - C_N^b) \tag{B.28}$$

where

$$k_x = k_{CR}[(k_{CR}/k_m) + (C_N^{eq} + C_N^i)^{-1}]^{-1} \tag{B.29}$$

Two extreme cases are

$$\frac{k_{CR}}{k_m} \gg (C_N^{eq} + C_N^i)^{-1}; \quad \frac{dC_N^b}{dt} = k_m(C_N^{eq} - C_N^b)$$

i.e. mass transport control; and

$$\frac{k_{CR}}{k_m} \ll (C_N^{eq} + C_N^i)^{-1}; \quad \frac{dC_N^b}{dt} = k_{CR}[(C_N^{eq})^2 - (C_N^b)^2]$$

i.e. chemical reaction control (second order).

The above model predicts mass transfer control at low surfactant concentration and chemical reaction control at high surfactant concentration.

B.8 Hydrodynamic model

This model takes into account the influence of hydrodynamic and surface tension factors on nitrogen adsorption and desorption; the mass transfer coefficient is given by

$$k = v_e^{3/2} \rho^{1/2} \sigma^{-1/2} D^{1/2} \tag{B.30}$$

where

v_e = characteristic eddy velocity

ρ = density

σ = surface tension

D = diffusivity of gas

When the surfactant film is dominant it resists surface renewal by eddies and the mass transfer coefficient k is modified as

$$k' = k(1 - \theta)^{1/2} \tag{B.31}$$

If $\theta = 1$, $k' = 0$ but experimentally a finite value of k' is obtained. The model had a limited success.

B.9 Mixed transport control model

The mass transfer in the metal and gas phases is assumed to be rate controlling ([14] in Chapter 3) and thermodynamic equilibrium is assumed to exist at the interface:

$$(N_2)_g = 2[N] \tag{B.32}$$

$$K = (f_N^i C_N^i)^2/(p_{N_2}^i/RT) \tag{B.33}$$

f_N^i is the activity coefficient, C_N^i is the interfacial concentration of nitrogen and $p_{N_2}^i$ is the corresponding partial pressure.

The flux (J_{N_2}) in the gas phase is given by

$$J_{N_2} = \frac{k_g}{RT}(p_{N_2} - p_{N_2}^i) \tag{B.34}$$

Similarly, in the metal phase,

$$J_m = k_m(C_N^i - C_N^b) \tag{B.35}$$

where C_N^b is the bulk metal concentration and k_m and k_g are the mass transfer coefficients in the metal and gas phases, respectively. Noting the equality of fluxes

$$J_{N_2} = \tfrac{1}{2}J_m \tag{B.36}$$

the interfacial terms $p_{N_2}^i$ and C_N^i can be eliminated through Equations (B.35) and (B.33) and the resulting flux equation can then be either numerically or analytically integrated to give nitrogen concentration as a function of time. The following expression is obtained after integration ([14] in Chapter 3):

$$\frac{(\alpha/2) + \%N_e}{\%N_e} \times \ln\left(\frac{\sqrt{[A - (\alpha/2) - \%N_e]}}{\sqrt{[B - (\alpha/2) - \%N_e]}}\right)$$

$$- \frac{(\alpha/2) - \%N_e}{\%N_e} \times \ln\left(\frac{\sqrt{[A - (\alpha/2) - \%N_e]}}{\sqrt{[B - (\alpha/2) - \%N_e]}}\right) = -k_m A_i t / V \tag{B.37}$$

where

$\%N_e$ = equilibrium nitrogen in metal (wt %)

α $\quad = k_m KRT\rho/2800k_g f_N^2$; ρ is density

A $\quad = \sqrt{[\alpha(\%N) + (\alpha/2)^2 + \%N_e^2]}$; $\%N$ is nitrogen wt % at any time t

B $\quad = \sqrt{[\alpha(\%N_o) + (\alpha/2)^2 + \%N_e^2]}$; $\%N_o$ is initial wt % nitrogen

A_i $\quad = $ nominal interface area

V $\quad = $ volume of metal

t $\quad = $ time

Equation (B.37) can be solved by the Newton–Raphson method to calculate $\%N$ for any given value of t. For the sake of simplicity only, it is assumed that f_N is constant. In the numerical integration procedure f_N can be varied with concentration after each step of the iteration.

Summary of rate equations for the oxidation of dissolved carbon by gaseous oxygen

The direct oxidation of carbon dissolved in metal can take place in the jet impact zone according to

$$[C] + \tfrac{1}{2}(O_2)_g = (CO)_g$$

which is the same as Equation (6.1).

Since oxygen is readily adsorbed by iron, the adsorption of oxygen as a rate-controlling step can be neglected. In the main blow period of constant dC_C/dt, since the bath has a high carbon content, the mass transfer of carbon as a rate-controlling step can be neglected. The possible rate-controlling steps are thus the following:

1. Mass transfer (counter current) of carbon monoxide and oxygen in the gas phase.
2. Interfacial chemical reaction between carbon and oxygen.

Rate equations proposed in different models reported in the literature ([10–15] in Chapter 6) are summarized below.

Rate equation assuming mass transfer control in the gas phase

The equation for the flux of oxygen (J_{O_2}) in the gas phase can be written as

$$J_{O_2} = -k_g\left(\frac{p}{RT_f}\right)\ln\left(\frac{1 + p_{O_2}^i/p}{1 + p_{O_2}^b/p}\right) \qquad \text{(C.1)}$$

where

    k_g = mass transfer coefficient in gas film

    p = total pressure

    T_f = average gas film temperature

    $p_{O_2}^i$ and $p_{O_2}^b$ = interfacial and bulk pressure of oxygen, respectively

The derivation of Equation (C.1) is similar to that of Equation (D.5) (discussed later). From mass balance the flux of carbon monoxide gas (J_{CO}) is twice that of oxygen:

$$J_{O_2} = -\tfrac{1}{2} J_{CO} \tag{C.2}$$

The rate of removal of carbon is expressed in terms of the flux of oxygen (for high-carbon melts only) as

$$-\frac{dC_C}{dt} = 200 \left(\frac{M_C A}{\rho V} \right) J_{O_2} \tag{C.3}$$

where C_C is wt % of carbon in the melt ($M_C = 12$ for the atomic weight of carbon) and ρ is the density of melt, V the volume of the melt and A the interfacial area. From Equations (C.1) and (C.3), if $p_{O_2}^i \ll 1$ and $p = 1$ atm, the rate equation is given by

$$-\frac{dC_C}{dt} = \frac{2400 A}{\rho V} \left(\frac{k_g}{RT_f} \right) \ln(1 + p_{O_2}^b) \tag{C.4}$$

Rate equation assuming interfacial chemical reaction control

The elementary steps in the chemical reaction will be:

Dissociative chemisorption	$(O_2^i)_g = 2[O^*]$	(C.5)
Surface adsorption	$[C] = [C^*]$	(C.6)
Chemical reaction	$[C^*] + [O^*] = (CO^*)_g$	(C.7)
Desorption	$[CO^*] = (CO^i)_g$	(C.8)

The symbol * indicates chemisorbed material. At high carbon contents, since dC_C/dt is constant and carbon is not surfactive, the steps (C.6) and (C.7) can be neglected as rate controlling. Also, since the rate of decarburization is independent of pressure of carbon monoxide ([12] in Chapter 6), Equation (C.8) is not rate controlling.

If the backward reaction in Equation (C.5) is neglected, the rate of decarburization can be written as

$$-\frac{dC_C}{dt} = \frac{2400 A}{\rho V} k_f p_{O_2}^i \tag{C.9}$$

where k_f is the forward reaction rate constant (mol/m²s atm). In the presence of other surfactive elements like sulphur, the overall rate (r_o) can be written as

$$r_o = (1 - \theta_s) r_u + \theta_s r_s \tag{C.10}$$

where θ_s is the fraction of sites covered by sulphur and r_u and r_s are respectively the rate through bare and covered sites. If K_S is the adsorption coefficient of sulphur, then

$$\theta_s = \frac{K_S f C_S}{1 + K_S f C_S} \tag{C.11}$$

where $f = \gamma_S^b / \gamma_S^*$, i.e. the ratio of the activity coefficients of sulphur in the bulk and at the interface, and C_S is the concentration of sulphur (wt %) in the bulk metal. The rates r_u and r_s are calculated as

$$r_u = \left(\frac{2400A}{\rho V}\right) k_{ru} p_{O_2}^i$$

$$r_s = \left(\frac{2400A}{\rho V}\right) k_{rs} p_{O_2}^i \qquad \text{(C.12)}$$

where k_{ru} and k_{rs} are the respective rate constants. From Equations (C.8)–(C.12) we obtain the expression for overall rate, r_o, as

$$r_o = -\frac{dC_C}{dt} = \left(\frac{2400A}{\rho V}\right) \left(\frac{k_{ru} + k_{rs} K_S f C_S}{1 + K_S f C_S}\right) p_{O_2}^i \qquad \text{(C.13)}$$

If decarburization is controlled by the interfacial reaction only, $p_{O_2}^i = p_{O_2}^b$. Also, if $k_{rs} = k_{ru}$ (i.e. in the absence of sulphur) Equation (C.13) reduces to the simple rate equation (C.9). Laboratory experiments have shown that decarburization is only negligibly retarded by sulphur ([13] in Chapter 6). Even the oxygen dissolution rate is not affected by sulphur ([14] in Chapter 6).

The role of sulphur in the oxygen steelmaking reactions, however, is not clear. Waste gas analysis of the converter has shown that about 6–8% of the sulphur in the metal is removed by direct oxidation in the hot spot zone. This would imply a high interfacial concentration of sulphur. It is difficult to quantify the influence of adsorbed sulphur at the interface on the rate of carbon oxidation in the oxygen steelmaking process. Further, Equation (C.4) is essentially of the same form as Equation (C.9) because in the former $\ln(1 + p_{O_2}^b) \simeq p_{O_2}^b$. In order to distinguish between mass transfer and chemical reaction the actual values of k_g and k_f at different temperatures must be accurately known, which is not an easy task.

Rate equation assuming mixed control

Considering the fact that dC_C/dt increases with oxygen flow rate in the main blow period, it is reasonable to postulate that the rate of carbon oxidation is controlled by mass transport in the gas phase as well as by chemical reaction at the interface.

With reference to Equation (C.1), if $p = 1$ atm and $\ln(1 + p_{O_2}^i) = p_{O_2}^i$, one obtains

$$-\frac{dC_O}{dt} = \frac{2400A}{\rho V} \left(\frac{k_g}{RT_f}\right) [\ln(1 + p_{O_2}^b) - p_{O_2}^i] \qquad \text{(C.14)}$$

$p_{O_2}^i$ in the above equation can be substituted from Equation (C.13) to give the rate equation for mixed control. It has been demonstrated that at high carbon contents the decarburization reaction may be controlled by the mass transfer of oxygen in the gas phase ([10, 11] in Chapter 6).

Summary of rate equations for the oxidation of carbon by CO/CO_2 gas mixture

The periphery of the oxygen jet is contaminated as a result of the entrainment of gases from the surrounding atmosphere. The carbon monoxide gas entrained in the jet is oxidized to carbon dioxide. The overall reaction of dissolved carbon with CO/CO_2 gas mixture can be written as

$$(CO_2)_g + [C] = (CO)_g$$

which is the same as Equation (6.2).

The above reaction involves the following steps ([13]–[18] in Chapter 6):

1. Mass transfer (counter current) of carbon monoxide and carbon dioxide in the gas phase.
2. Mass transfer of carbon in the metal phase.
3. Chemical reaction at the interface.

At high carbon contents and constant dC_C/dt, step 2 can be neglected as a rate-controlling step.

Rate equations assuming mass transfer control in the gas phase only

In a binary gas mixture of carbon monoxide and carbon dioxide the steady-state flux of carbon dioxide (J_{CO_2}) across the stagnant film can be written as

$$J_{CO_2} = -\frac{D_{CO_2\text{-}CO}}{RT_f}\frac{dp_{CO_2}}{dz} + (J_{CO_2} + J_{CO})\frac{p_{CO_2}}{p} \tag{D.1}$$

where

$\quad J_{CO}$ = flux of carbon monoxide

$\quad z$ = position coordinate in film

$\quad D_{CO_2\text{-}CO}$ = binary diffusivity of carbon dioxide and carbon monoxide

p_{CO_2}, p, T_f and R have the same significance as before. Balancing the fluxes of carbon monoxide and carbon dioxide:

$$J_{CO} = -2J_{CO_2} \tag{D.2}$$

On substituting into Equation (D.1)

$$J_{CO_2} = -\frac{D_{CO_2-CO}}{RT_f} p \frac{d[\ln(p + p_{CO_2})]}{dz} \tag{D.3}$$

If the gas film thickness is δ, then integration over this thickness yields

$$J_{CO_2} = -p \frac{D_{CO_2-CO}}{RT_f\delta} \ln\left(\frac{1 + p_{CO_2}^i/p}{1 + p_{CO_2}^b/p}\right) \tag{D.4}$$

If $D_{CO_2-CO}/\delta = k_g'$, the mass transfer coefficient

$$J_{CO_2} = -k_g'(p/RT_f)\ln\left(\frac{1 + p_{CO_2}^i/p}{1 + p_{CO_2}^b/p}\right) \tag{D.5}$$

This equation is similar to Equation (C.1). If $p_{CO_2}^i \to 0$ and $p = 1$ atm, the rate equation at high carbon contents as in the case of Equation (C.4) is given by

$$-\frac{dC_C}{dt} = \frac{1200A}{\rho V} \frac{k_g'}{RT_f} \ln(1 + p_{CO_2}^b) \tag{D.6}$$

Rate equation assuming interfacial chemical reaction control

The essential steps of the chemical reaction can be written as:

Adsorption and dissociation	$[CO_2^i] = [CO^*] + [O^*]$	(D.7)
Surface adsorption	$[C] = [C^*]$	(D.8)
Chemical reaction	$[C^*] + (O^*) = (CO^*)$	(D.9)
Desorption	$[CO^*] = [CO^i]$	(D.10)

Since carbon is not surfactive and at high carbon contents dC_C/dt is independent of carbon content, Equation (D.8) can be neglected as a rate-controlling step. The intrinsic rate of equation (D.9) is known to be faster than that of Equation (D.7). Also, since the rate of decarburization is independent of p_{CO}, both Equations (D.9) and (D.10) can be neglected as rate-controlling steps. Thus, dissociative adsorption of carbon dioxide is likely to be the rate-controlling step. If the backward reaction in Equation (D.7) is neglected, then

$$-\frac{dC_C}{dt} = \frac{1200A}{\rho V} k_r'(p_{CO_2}^i - p_{CO_2}^e) \tag{D.11}$$

where $p_{CO_2}^e$, the equilibrium pressure, is small and can be neglected. Hence

$$-\frac{dC_C}{dt} = \frac{1200A}{\rho V} k_r' p_{CO_2}^i \tag{D.12}$$

where k_r' is the rate constant. Equation (D.12) is similar to Equation (C.9).

Rate equation assuming mixed control

The $p_{CO_2}^i$ from Equation (D.12) can be substituted into Equation (D.5) to obtain the rate equation for mixed control.

Rate equation assuming mixed mass transport control in metal and gas

In order to derive the rate equations for decarburization by CO/CO_2 at low concentrations of carbon (i.e. below C_{crit} when the mass transfer in the metal can become rate controlling; see also Chapter 7 for decarburization models in low-carbon steels), the chemical reactions at the interface can be assumed to be infinitely fast, i.e. thermodynamic equilibrium is assumed to exist at the interface ([17] in Chapter 6). Only two of the following three reactions need to be considered:

$$[O] + (CO)_g = (CO_2)_g \tag{D.13}$$

$$[C] + (CO_2)_g = 2(CO)_g \tag{D.14}$$

$$[C] + [O] = (CO)_g \tag{D.15}$$

If the respective interfacial and bulk concentrations are denoted by the superscripts i and b, the fluxes of carbon monoxide and carbon dioxide in Equation (D.13) can be written as

$$J_{CO_2} = \frac{k_g}{RT} (p_{CO_2}^b - p_{CO_2}^i) \tag{D.16}$$

$$J_{CO} = \frac{k_g}{RT} (p_{CO}^b - p_{CO}^i) \tag{D.17}$$

Similarly, for Equation (D.15):

$$J_C = k_C(C_C^i - C_C^b) \tag{D.18}$$

$$J_O = k_O(C_O^i - C_O^b) \tag{D.19}$$

where k_g, k_C and k_O are the mass transfer coefficients, and subscripts C and O indicate the concentrations of carbon and oxygen in the metal. From the equality of fluxes

$$J_O = J_{CO} + J_{CO_2} \tag{D.20}$$

$$J_C = J_{CO} + J_{CO_2} \tag{D.21}$$

Considering thermodynamic equilibrium at the interface

$$\frac{p^i_{CO_2}}{p^i_{CO} C^i_O} = K_1 \qquad\qquad\qquad \text{(D.22)}$$

$$\frac{p^i_{CO}}{C^i_C C^i_O} = K_2 \qquad\qquad\qquad \text{(D.23)}$$

The Equations (D.18)–(D.19) and (D.22)–(D.23) constitute four equations with four unknown interfacial concentrations, p^i_{CO}, $p^i_{CO_2}$, C^i_C and C^i_O, and can be solved by an iterative procedure. The carbon removed at each time step can now be calculated from Equation (D.18) (see Example 6.1).

Although the oxidation of dissolved carbon has been a subject of great theoretical interest and several rate equations have been proposed, the actual process of the oxidation of carbon in a converter is very complex and difficult to model unless three-phase slag–metal–gas reactions are considered simultaneously (see Example 6.1).

Estimation of effective area factor for a partially submerged particle m

The value of the effective area factor for a partially submerged particle m can be estimated as follows. The volume of the submerged portion of the particle is given by

$$V_{sub} = \frac{\pi}{32} d_p^3 [3(1 - \cos \theta_o) - \tfrac{1}{3}(1 - \cos 3\theta_o)]$$ (E.1)

where

d_p = diameter of the particle

V_{sub} = volume of the submerged portion

θ_o = half of the angle subtended by the submerged portion at the centre

The surface area in contact with the melt is given A_{sub} by

$$A_{sub} = \frac{\pi d_p^2}{2} (1 - \cos \theta_o)$$ (E.2)

Using Archimedes' principle, i.e. equating the weight of the liquid displaced by the weight of the particle,

$$V_{sub} = \frac{\rho_s}{\rho_m} \left(\frac{\pi}{6} d_p^3 \right)$$ (E.3)

where ρ_s is the density of the slag-type particle (2000 kg/m^3 for calcium carbide) and ρ_m is the density of metal (6900 kg/m^3). Thus we get

$$A_{sub} = 1.1 d_p$$ (E.4)

The total volume of the PIB is given by V_p^g. Thus the total surface area of particles in bubble PIB in contact with the metal is given by

$$A_{tot} = \frac{V_p^g}{(\pi/6)d_p^3} A_{sub}$$ (E.5)

From Equations (E.4) and (E.5) we get

$$A_{tot} = 2.1V_p^g/d_p \tag{E.6}$$

On comparing Equation (E.6) with Equation (8.17) the value of m can be estimated.

It has already been pointed out that there is a limit on the maximum value of m owing to a limit on the bubble base area available for desulphurization. This value of m, denoted as m_{max}, may be estimated as follows.

The total number of particles sitting on the base of a bubble, when the base is closely packed by the particles, is given by

$$n_{p,max} = \frac{(\pi/4)x^2}{(\pi/4)d_p^2} \tag{E.7}$$

where x is the diameter of the hemispherical cap bubble. The area available for each particle is A_{sub}. Hence the maximum area available is given by

$$A_{tot,max} = n_{p,max}A_{sub}n_b \tag{E.8}$$

where n_b is the number of bubbles. Comparison of Equation (E.8) with Equation (8.17) gives

$$m_{max} = 4/\pi \times 1.1 = 1.4$$

In Example 8.1 a value of $m = 1$ has been used in computations.

Index